D1244067

Real Algebraic Differential Topology
Part I

Richard S. Palais

Brandeis University

WILLIAM MADISON RANDALL LIBRARY UNC AT WILMINGTON

COPYRIGHT © PUBLISH OR PERISH, INC. 1981
All rights reserved.

ISBN 0-914098-19-5
Library of Congress Catalog Card Number: 81-81990

PUBLISH OR PERISH, INC.
901 WASHINGTON STREET
WILMINGTON, DE 19801 (U.S.A.)

In Japan distributed exclusively by
KINOKUNIYA COMPANY, LTD.
TOKYO

QA612
.P295
pt. 1

Introduction.

A real algebraic variety, naively at least, is a subset V of $\mathbb{R}^n$ consisting of the common zeros of some set I of polynomials in n variables with real coefficients. We may as well assume that I is the ideal in $\mathbb{R}[X_1, \ldots, X_n]$ of all polynomials vanishing on V, and we say that the variety is non-singular and of dimension k when for each of its points there is a neighborhood U in $\mathbb{R}^n$ and polynomials $Y_1, \ldots, Y_n$, forming a local coordinate system in U, such that $Y_{k+1}, \ldots, Y_n$ belong to I, and in fact $V \cap U = \mathcal{O}$ is the subset of U where $Y_{k+1}, \ldots, Y_n$ simultaneously vanish. In this case V is a closed, regularly embedded, real analytic submanifold of $\mathbb{R}^n$ and $Y_1, \ldots, Y_k$ is a local coordinate system for V in $\mathcal{O}$.

Non-singular real algebraic varieties lie at the interface of two great modern mathematical theories, differential topology and algebraic geometry, and it is only a mild distortion to claim that both theories had their origins in the study of these objects. Be that as it may, in recent years the two subjects have developed in very different directions.

In algebraic geometry there has been a series of extensive generalizations. From varieties defined over the real and complex numbers to those defined over arbitrary ground fields of any characteristic. From affine varieties to projective varieties to abstract varieties and eventually to schemes. Partly as a result of this trend it often seems that there is now little in common to the two fields.

Still, the simple but beautiful question of what the relation is between smooth manifolds and smooth varieties has not been entirely neglected. It

271788

has been studied by, among others, Seifert, Whitney, Nash, Wallace, and Tognoli, and we now know for example that any compact C^1 manifold M is diffeomorphic to a non-singular real algebraic variety. We even know that if a compact Lie group G acts smoothly on M then an algebraic structure can be picked for M so that the action of G (which has a unique structure of real algebraic group) is algebraic. It is with questions of this nature that we shall be ultimately concerned in what follows. But our more general goal is to develop a setting for studying "real algebraic differential topology" which is hopefully well suited to the subject.

The current attempt at this exposition is in fact version number four. Originally my view of varieties was the naive one suggested by the definition above; they were always subvarieties of some $\mathbb{R}^n$, and morphisms between them were restrictions of polynomial maps between their ambient spaces. Gradually my algebraic geometer friends have educated and prodded me into accepting the more elegant and intrinsic ringed space point of view and at least a little of the yoga of schemes. I have tried to be selective and conservative in this transition, for there is strong evidence that from the algebraic geometer's viewpoint there is something particularly simple about real algebraic varieties. The full power (and complexity) of the scheme machinery is wasted, and even the more sophisticated version of a "ringed space" structure (where one uses a sheaf of rings rather than the global ring of sections) is unnecessary. It has been my goal to exploit this simplicity, to adopt and adapt that part of the modern approach to algebraic geometry that is useful for exploring its relation to differential topology and avoid those parts which add great generality to the theory but are not apparently relevant to studying differentiable manifolds.

A question we should mention here (because we shall not come back to it later) is the obvious one, why try to give a smooth manifold the structure of an algebraic variety. I will suggest only one possible response, namely that a real algebraic variety is inherently a much simpler object than a smooth manifold; it can for example be given by a "finite amount of data". If one were to try to deal algorithmically with smooth manifolds, even in a theoretical way, it would be important to be able to specify them in such a concrete finitistic way.

ACKNOWLEDGEMENTS

In learning what I needed to write what follows, I must have asked ten thousand questions of my many friends versed in algebraic geometry. More than a few of these questions must have at least seemed foolish. Since there are too many of you to thank individually, I express my appreciation here to you all for your patience.

To Mike Artin my special thanks for the many times he helped me past serious obstacles.

The final manuscript was prepared while I was on sabbatical leave from Brandeis at the Institute for Advanced Study in Princeton. I would like to thank these institutions for supporting my research and also the National Science Foundation which partially supported my research through grant GP-28376.

CONTENTS

1. Preliminaries.

2. Affine Algebraic Geometry

1.0. Notations and Conventions.

In what follows K will denote some field which is arbitrary but is supposed fixed throughout the local context. By an algebra we shall mean a commutative algebra over K, with unit. To avoid endless repetition of "Let $\mathcal{A}$ be an algebra..." we adopt the convention that whenever "$\mathcal{A}$" appears in a discussion without being formally introduced, it denotes an algebra.

We shall regard the ground field K as a subalgebra of every algebra $\mathcal{A}$ via the standard identification; i.e., $\lambda \in K$ is identified with λ times the identity element of $\mathcal{A}$. Thus the identity 1 of K is also the identity of $\mathcal{A}$. Since our algebras will most frequently be algebras of K valued functions on some set S this means we are identifying $\lambda \in K$ with the constant function on S with values λ at every point. A homomorphism between two algebras is always assumed to be "unitary", i.e., to map the identity element to the identity element. Thus if $h : \mathcal{A}_1 \to \mathcal{A}_2$ is a homomorphism of algebras, then by the above conventions $K \subseteq \mathcal{A}_1 \cap \mathcal{A}_2$ and h restricted to K is the identity map. In particular, a homomorphism $h : \mathcal{A} \to K$ is a homomorphic "retraction" of $\mathcal{A}$ onto K.

If $\mathcal{A}$ is an algebra then $\mathcal{A}^*$ will denote its dual as a vector space over K, i.e., all linear maps $\mathcal{A} \to K$, and $\hat{\mathcal{A}} \subseteq \mathcal{A}^*$ will denote its "dual" as an algebra over K, i.e., all homomorphisms $\mathcal{A} \to K$. (If $\mathcal{A}$ is represented by a complex symbol, we may write $\mathcal{A}^\wedge$ instead of $\hat{\mathcal{A}}$). Given a homomorphism $h : \mathcal{A}_1 \to \mathcal{A}_2$ of algebras then h is in particular linear and so induces $h^* : \mathcal{A}_2^* \to \mathcal{A}_1^*$ defined by $\ell \mapsto \ell \circ h$. Moreover h^* restricts to a map $\hat{h} : \hat{\mathcal{A}}_2 \to \hat{\mathcal{A}}_1$. It is trivial that if h is surjective then $\hat{h}$ is injective. (It is not true that if h is injective then $\hat{h}$ is necessarily surjective; consider the inclusion of K in an extension field.)

Given a set S we denote by K^S the algebra of all K-valued functions on S with pointwise operations. We note that there is a canonical map of S into $(K^S)^\wedge$ called "evaluation" and denoted by $Ev : S \to (K^S)^\wedge$. Namely, $Ev(s) : K^S \to K$ is defined by $f \mapsto f(s)$.

The map $S \mapsto K^S$ is a contravariant functor from the category of sets to the category of algebras. Given $f : S_1 \to S_2$ the induced homomorphism $K^f : K^{S_2} \to K^{S_1}$ is given by $g \mapsto g \circ f$. We will sometimes write f^* for this homomorphism. We note that if f is surjective then K^f is injective and if f is injective K^f is surjective.

1.1. Representiations of Algebras as Function Algebras.

1.1.1. Definition. A representation of $\mathcal{A}$ is a homomorphism $\rho : \mathcal{A} \to K^S$ (where S is some set) such that given distinct points s_1 and s_2 of S there is an $x \in \mathcal{A}$ such that $\rho(x)(s_1) \neq \rho(x)(s_2)$.

1.1.2. Remark. If the stated condition is not met then we can still get a representation $\tilde{\rho} : \mathcal{A} \to K^{\tilde{S}}$ where $\tilde{S} = S/\sim$ is the quotient set of S under the equivalence relation $s_1 \sim s_2$ if $\rho(x)(s_1) = \rho(x)(s_2)$ for all $x \in \mathcal{A}$.

1.1.3. Remark. Given a representation $\rho : \mathcal{A} \to K^S$ we have an associated map $s \mapsto \varphi_s$ of S into $\hat{\mathcal{A}}$, namely, $\varphi_s(x) = \rho(x)(s)$. The condition of 1.1.1 is just that this map be injective, so that we can identify S naturally with a subset of $\hat{\mathcal{A}}$. Note that $s \mapsto \varphi_s$ is the composition of $\hat{\rho} : (K^S)^\wedge \to \hat{\mathcal{A}}$ and $Ev : S \to (K^S)^\wedge$.

1.1.4. Remark. If $T \subseteq S$ then the inclusion map $i_T : T \to S$ induces a surjection $K^{i_T} : K^S \to K^T$, namely $g \mapsto g \circ i_T = g|T$. If $\rho : \mathcal{A} \to K^S$ is a representation of $\mathcal{A}$ then so is $K^{i_T} \circ \rho : \mathcal{A} \to K^T$.

1.1.5. Definition. If $\rho : \mathcal{A} \to K^S$ is a representation of $\mathcal{A}$ then for each subset T of S we define a representation $\rho_{|T} : \mathcal{A} \to K^T$ of $\mathcal{A}$, called the subrepresentation of $\mathcal{A}$ defined by T, by $\rho_{|T}(x) = \rho(x)|T$.

1.1.6. Definition. Two representations $\rho_1 : \mathcal{A} \to K^{S_1}$ and $\rho_2 : \mathcal{A} \to K^{S_2}$ are equivalent if there exists a bijection $f : S_1 \to S_2$ such that $\rho_2 = K^f \circ \rho_1$.

1.1.7. Exercise. Let $\rho_1 : \mathcal{A} \to K^{S_1}$ and $\rho_2 : \mathcal{A} \to K^{S_2}$ be two representations of $\mathcal{A}$. Suppose there exists a map $f : S_2 \to S_1$ such that $\rho_2 = K^f \circ \rho_1$. Show that f is uniquely determined and that f is necessarily injective. In particular if $\rho_1 = \rho_2$ then f must be the identity map of S_1. Deduce that if there also exists a map $g : S_1 \to S_2$ such that $\rho_1 = K^g \circ \rho_2$ then f and g are uniquely determined inverse bijections of S_1 with S_2, so that ρ_1 and ρ_2 are equivalent.

1.1.8. Proposition. Let $\rho : \mathcal{A} \to K^S$ and $\bar{\rho} : \mathcal{A} \to K^{\bar{S}}$ be two representations of $\mathcal{A}$. A necessary and sufficient condition that $\bar{\rho}$ be equivalent to a subrepresentation of ρ is that there exist a map $f : \bar{S} \to S$ such that $\bar{\rho} = K^f \circ \rho$. This map if it exists is unique. In particular $T = \mathrm{im}(f)$ is unique.

Proof. Immediate from 1.1.7. ■

1.1.9. Definition. A representation $\rho : \mathcal{A} \to K^S$ is called a universal representation of $\mathcal{A}$ if every representation of $\mathcal{A}$ is equivalent to a subrepresentation of ρ.

1.1.10. Remark. By 1.1.8, if $\rho : \mathcal{A} \to K^S$ is universal then every representation of $\mathcal{A}$ is equivalent to a unique subrepresentation of ρ.

1.1.11. <u>Exercise.</u> Show that two universal representations of $\mathcal{A}$ are equivalent.

1.1.12. <u>Definition.</u> We define a representation $\rho^{\mathcal{A}} : \mathcal{A} \to K^{\hat{\mathcal{A}}}$ of $\mathcal{A}$, called the Gelfand representation of $\mathcal{A}$, by $\rho^{\mathcal{A}}(x)(\varphi) = \varphi(x)$.

1.1.13. <u>Exercise.</u> Check that the Gelfand representation of $\mathcal{A}$ is in fact a representation.

1.1.14. <u>Notation.</u> Because of its unique importance we adopt a special notation for the Gelfand representation of $\mathcal{A}$. Given $x \in \mathcal{A}$ we write $\hat{x} = \rho^{\mathcal{A}}(x) \in K^{\hat{\mathcal{A}}}$. Thus $\hat{x}$ is the function $\hat{\mathcal{A}} \to K$ defined by $\hat{x}(\varphi) = \varphi(x)$.

1.1.15. <u>Theorem.</u> The Gelfand representation of $\mathcal{A}$ is universal.

<u>Proof.</u> We have seen in 1.1.3 that given a representation $\rho : \mathcal{A} \to K^S$ we have a natural map $s \mapsto \varphi_s$ of S into $\hat{\mathcal{A}}$ defined by $\varphi_s(x) = \rho(x)(s)$. Calling this map $f : S \to \hat{\mathcal{A}}$, we have an induced map $K^f : K^{\hat{\mathcal{A}}} \to K^S$ and for all $s \in S$ we have:
$(K^f \circ \rho^{\mathcal{A}}(x))(s) = (K^f \circ \hat{x})(s) = \hat{x} \circ f(x) = \hat{x}(\varphi_s) = \varphi_s(x) = \rho(x)(s)$; i.e., we have $K^f \circ \rho^{\mathcal{A}} = \rho$ and by 1.1.8 ρ is equivalent to a subrepresentation of $\rho^{\mathcal{A}}$. ■

1.1.16. <u>Proposition.</u> The Gelfand representation is natural in the sense that if $h : \mathcal{A}_1 \to \mathcal{A}_2$ is a homomorphism of algebras then the following diagram commutes.

$$\begin{array}{ccc} \mathcal{A}_1 & \xrightarrow{\;h\;} & \mathcal{A}_2 \\ {\scriptstyle \rho_{\mathcal{A}_1}}\big\downarrow & & \big\downarrow{\scriptstyle \rho_{\mathcal{A}_2}} \\ K^{\hat{\mathcal{A}}_1} & \xrightarrow{\;K^{\hat{h}}\;} & K^{\hat{\mathcal{A}}_2} \end{array}$$

Proof. Given $x \in \mathcal{A}_1$ we must show that $(h(x))^\wedge = \hat{x} \circ \hat{h}$; i.e., for each $\varphi \in \hat{\mathcal{A}}_2$ we must show that $(h(x))^\wedge(\varphi) = \hat{x}(\hat{h}(\varphi))$ or that $\varphi(h(x)) = (\hat{h}(\varphi))(x)$. But by definition of $\hat{h} : \hat{\mathcal{A}}_1 \to \hat{\mathcal{A}}_2$, $\hat{h}(\varphi) = \varphi \circ h$. ■

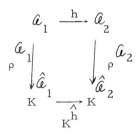

1.2. Strict Semi-Simplicity.

1.2.1. Definition. An ideal $\mathcal{I}$ in the algebra $\mathcal{A}$ over K is called strictly maximal if it has codimension one in $\mathcal{A}$. We denote the set of all strictly maximal ideals of $\mathcal{A}$ by $\mathrm{Spec}(\mathcal{A})$ (or by $\mathrm{Spec}_K(\mathcal{A})$ if the identity of ground field K is ambiguous).

1.2.2. Remark. The notation $\mathrm{Spec}(\mathcal{A})$ is sometimes used to denote the set of all maximal ideals of $\mathcal{A}$ and sometimes to denote the set of all prime ideals of $\mathcal{A}$.

1.2.3. Proposition. The map $\varphi \mapsto \ker(\varphi)$ is a bijective correspondence of $\hat{\mathcal{A}}$ with $\mathrm{Spec}(\mathcal{A})$.

Proof. If $\varphi \in \hat{\mathcal{A}}$ then in particular φ is a surjective linear map $\mathcal{A} \to K$ so $\ker(\varphi)$ has codimension one in $\mathcal{A}$. Conversely if $M \in \mathrm{Spec}(\mathcal{A})$ then $\mathcal{A}/M$ is a one-dimensional algebra over K (which has an identity, namely, $1+M$) so there is a unique homomorphism $\psi : \mathcal{A}/M \to K$, which is of course an isomorphism. If $\prod : \mathcal{A} \to \mathcal{A}/M$ is the canonical projection then $\varphi = \psi \circ \prod$ is the unique element of $\hat{\mathcal{A}}$ with kernel M. ■

1.2.4. Remark. In view of 1.2.3 the sets $\hat{\mathcal{A}}$ and $\mathrm{Spec}(\mathcal{A})$ can be used more or less interchangeably. For example, the Gelfand representation could just as well be thought of as representing $\mathcal{A}$ as functions on

$\mathrm{Spec}(\mathcal{A})$. If $x \in \mathcal{A}$ the value of $\hat{x}$ at $M \in \mathrm{Spec}(\mathcal{A})$ is defined by the condition $x-\hat{x}(M) \in M$. We shall tend to "prefer" $\hat{\mathcal{A}}$ over $\mathrm{Spec}(\mathcal{A})$; however, we will frequently remark on the appropriate changes to be made if one prefers to work with $\mathrm{Spec}(\mathcal{A})$.

1.2.5. Definition. The kernel of the Gelfand representation $\rho^{\mathcal{A}} : \mathcal{A} \to K^{\hat{\mathcal{A}}}$ will be denoted by $\mathrm{Rad}_{\mathcal{A}}(0)$ and called the strict radical of $\mathcal{A}$. We shall call $\mathcal{A}$ strictly semi-simple if $\mathrm{Rad}_{\mathcal{A}}(0) = (0)$; i.e., if the Gelfand representation of $\mathcal{A}$ is faithful.

1.2.6. Proposition. The strict radical of $\mathcal{A}$ is the intersection of all the strict maximal ideals of $\mathcal{A}$. That is

$$\begin{aligned} \mathrm{Rad}_{\mathcal{A}}(0) &= \cap \{M \mid M \in \mathrm{Spec}(\mathcal{A})\} \\ &= \cap \{\ker \varphi \mid \varphi \in \hat{\mathcal{A}}\}. \end{aligned}$$

Proof. By definition of $\mathrm{Rad}_{\mathcal{A}}(0)$

$$\begin{aligned} x \in \mathrm{Rad}_{\mathcal{A}}(0) &\Longleftrightarrow \hat{x} = 0 \\ &\Longleftrightarrow \hat{x}(\varphi) = 0 \quad \text{all } \varphi \in \hat{\mathcal{A}} \\ &\Longleftrightarrow \varphi(x) = 0 \quad \text{all } \varphi \in \hat{\mathcal{A}} \\ &\Longleftrightarrow x \in \ker(\varphi) \quad \text{all } \varphi \in \hat{\mathcal{A}}. \end{aligned}$$

■

1.2.7. Remark. The nil radical of $\mathcal{A}$ is defined as the intersection of all the maximal ideals of $\mathcal{A}$, and $\mathcal{A}$ is called semi-simple of its nil radical is zero. Clearly the nil radical is at least as small as the strict radical, but $\mathcal{A}$ can be semi-simple without being strictly semi-simple. For example, if $\mathcal{A}$ is a proper field extension of K then $\mathcal{A}$ has exactly one maximal ideal, namely, (0), so the nil radical of $\mathcal{A}$ is 0 and $\mathcal{A}$ is semi-simple. But $\mathcal{A}$ has no strictly maximal ideals, so $\mathrm{Rad}_{\mathcal{A}}(0) = \mathcal{A}$; i.e., the Gelfand representation of $\mathcal{A}$ is trivial.

1.2.8. Caution. As the above example clearly shows, the notions of strictly maximal ideal, strict radical, and strict semi-simplicity are relative to the choice of ground field K. For if $\mathcal{A}$ is an extension field of K and we think of $\mathcal{A}$ rather than K as our ground field, then of course (0) is strictly maximal, and $\mathcal{A}$ is strictly semi-simple.

1.2.9. Proposition. $\mathcal{A}$ is strictly semi-simple if and only if it admits a faithful representation $\rho : \mathcal{A} \to K^S$. Thus if ρ is any (not necessarily faithful) representation of $\mathcal{A}$ then $\mathcal{A}/\ker(\rho)$ is strictly semi-simple. In particular $\mathcal{A}/\mathrm{Rad}_{\mathcal{A}}(0)$ is always strictly semi-simple.

Proof. If $\mathcal{A}$ is strictly semi-simple then, by definition, the

Gelfand representation of $\mathcal{A}$ is faithful. If conversely $\mathcal{A}$ admits a faithful representation ρ then, since the Gelfand representation is universal, we can suppose ρ is the subrepresentation of the Gelfand representation defined by some subset $T \subseteq \hat{\mathcal{A}}$. If $x \in \mathcal{A}$ then $\hat{x}|T = 0$ implies $\hat{x} = 0$. A fortiori $\hat{x} = 0$ implies $x = 0$. ■

1.2.10. Example. Let K[X] denote as usual the algebra of polynomials in one variable with coefficients in K. Let $\rho : K[X] \rightarrow K^K$ denote the "standard" representation of K[X] as K-valued functions on K; i.e., for $f(X) \in K[X]$ and $\alpha \in K$, $\rho(f(X))(\alpha) = f(\alpha)$. The image of ρ will be denoted by $\mathcal{P}(K)$ and called the ring of polynomial functions on K. (Note by the way that the condition of 1.1.1 is clearly met. Given $\alpha_1 \neq \alpha_2$ in K let $f(X) = X-\alpha_1$; then $\rho(f(X))(\alpha_1) = 0$ while $\rho(f(X))(\alpha_2) = \alpha_2-\alpha_1 \neq 0$.) Now we know that, on general principles, the map which sends $\alpha \in K$ to the homomorphism $f(X) \mapsto f(\alpha)$ of K[X] onto K is an injective map $K \rightarrow K[X]^\wedge$ (cf. 1.1.3). It is in fact bijective. For suppose $\varphi \in K[X]^\wedge$ and let $\varphi(X) = \alpha$. If $f(X) = a_0 + a_1X + \ldots + a_nX^n$ then since φ is a homomorphism of algebras over K, $\varphi(f(X)) = a_0 + a_1\varphi(X) + \ldots + a_n\varphi(X)^n = f(\alpha)$. Henceforth we will use this canonical bijection to identify $K[X]^\wedge$ with K, so that ρ becomes the Gelfand representation of K[X]; i.e., $f(X)^\wedge(\alpha) = f(\alpha)$, or symbolically, $f(X)^\wedge = f$. It is now immediate that a sufficient condition for K[X] to be strictly semi-simple is that K be infinite, since a non-zero polynomial f(X) has a degree n and can have at most n roots. Moreover this sufficient condition is also necessary, for if $K = \{\alpha_1, \ldots, \alpha_n\}$ then defining $Q(X) = (X-\alpha_1)(X-\alpha_2)\ldots(X-\alpha_n)$,

$Q(X)$ is a polynomial of degree n, and in particular $Q(X) \neq 0$, and $Q(X)$ vanishes at each point of $K = K[X]^{\wedge}$. Indeed it is clear that the strict radical of $K[X]$ is just the principal ideal $(Q(X))$ generated by $Q(X)$. For $f(X)$ vanishes at $\alpha \in K$ if and only if $(X-\alpha)$ divides f and hence $f(X)$ vanishes identically on K if and only if it is divisible by all the $(X-\alpha_i)$ and hence by $Q(X)$. In the case that K is finite we note that the function $f_{\alpha_i} : K \to K$ defined by $f_{\alpha_i}(\alpha) = 0$ for $\alpha \neq \alpha_i$ and $f_{\alpha_i}(\alpha_i) = 1$ is in $\mathcal{P}(K)$. It is in fact clearly the image under ρ of the polynomial $f(X) = c \prod_{j \neq i}(X-\alpha_j)$ where $c = \prod_{j\neq i}(\alpha_i - \alpha_j)^{-1}$. Since the f_{α_i} for $i = 1, 2, \ldots, n$ form a basis for K^K over K it follows of course that $\mathcal{P}(K) = K^K$. On the contrary when K is infinite f_α is not in $\mathcal{P}(K)$ since it has infinitely many roots. For later reference we collect the above remarks as a proposition.

1.2.11. <u>Proposition</u>. The map $K[X]^{\wedge} \to K$ which sends φ to $\varphi(X)$ is a bijection. The inverse map associates to $\alpha \in K$ the homomorphism $f(X) \mapsto f(\alpha)$ of $K[X]$ onto K. Regarding the above bijection as an identification, the Gelfand representation of $K[X]$ is just the "usual" representation which associates to the formal polynomial $f(X) \in K[X]$ the polynomial function $K \to K$ defined by $\alpha \mapsto f(\alpha)$. The set of all such polynomial functions (i.e., the image of the Gelfand representation) is denoted by $\mathcal{P}(K)$ and called the algebra of polynomial function on K. If K is infinite then $K[X]$ is strictly semi-simple and $\mathcal{P}(K)$ is a proper subalgebra of K^K. On the contrary if K is finite then $K[X]$ is not strictly semi-simple; its strict radical in fact is the principal ideal generated by $Q(X) = \prod_{\alpha \in K} (X-\alpha)$, but in this case $\mathcal{P}(K)$ is all of K^K.

Proof. See 1.2.10 above. ■

1.2.12. Example. We now generalize the example 1.2.10 above. Let Γ be an abelian monoid (i.e., a commutative semi-group with unit). We shall define an algebra $K\{\Gamma\}$ called the semi-group algebra of Γ over K. As a vector space $K\{\Gamma\}$ consists of all functions $f : \Gamma \to K$ which vanish except at a finite number of points. If we identify $\gamma \in \Gamma$ with the function which is 1 at γ and 0 everywhere else in Γ, then clearly $\Gamma \subseteq K\{\Gamma\}$ and in fact Γ is a basis for $K\{\Gamma\}$. The general element f of $K\{\Gamma\}$ can now be written $\sum_{\gamma \in \Gamma} f(\gamma)\gamma$. The multiplication law in $K\{\Gamma\}$ is defined by requiring that two elements in Γ should have the same product in $K[\Gamma]$ as they do in Γ. Thus the product of $f = \sum_{\gamma_1} f(\gamma_1)\gamma_1$ and $g = \sum_{\gamma_2} g(\gamma_2)\gamma_2$ is

$$f*g = \sum_{\gamma_1, \gamma_2} f(\gamma_1)g(\gamma_2)\gamma_1\gamma_2$$

or "collecting terms"

$$f*g = \sum_{\gamma} \Big(\sum_{\gamma_1\gamma_2=\gamma} f(\gamma_1)g(\gamma_2)\Big)\gamma$$

or finally, in terms of functions, $f*g(\gamma) = \sum_{\gamma_1\gamma_2=\gamma} f(\gamma_1)g(\gamma_2)$. [In case Γ is a group, so that for each γ, γ_1 in Γ there is a unique γ_2 in Γ such that $\gamma_1\gamma_2 = \gamma$ (namely, $\gamma_2 = \gamma\gamma_1^{-1}$) we get back to the usual formula for the convolution product of the group algebra, namely, $f*g(\gamma) = \sum_{\gamma_1} f(\gamma_1)g(\gamma\gamma_1^{-1})$.] We note that the identity element of Γ is also the identity of $K\{\Gamma\}$ and we denote it by 1. There is an obvious but nevertheless important universal property of $K\{\Gamma\}$ (which in fact characterizes it up to unique isomorphism). Namely, every multiplicative homomorphism h of Γ into a K-algebra $\mathcal{A}$ extends uniquely to a homomorphism of the K-algebra $K\{\Gamma\}$ into $\mathcal{A}$. The extension is of course given by $h(\sum_\gamma f(\gamma)\gamma) = \sum_\gamma f(\gamma)h(\gamma)$; the definition of multiplication in $K\{\Gamma\}$ guarantees that this extension is a homomorphism of K-algebras, while the fact that Γ is a basis for $K\{\Gamma\}$ guarantees uniqueness. Using this property, it is easy to identify the dual object $K\{\Gamma\}^\wedge$. Namely, let $\hat{\Gamma}$ denote the dual semi-group to Γ (with respect to K), i.e., the semi-group of all "characters" $h : \Gamma \to K$, that is, multiplicative maps of Γ to K such that $h(1) = 1$. By the above extension property every such h extends uniquely to an element of $K\{\Gamma\}^\wedge$. Conversely, every element of $K\{\Gamma\}^\wedge$ restricts to an element of $\hat{\Gamma}$, so restriction is a bijective map of $K\{\Gamma\}^\wedge$ with $\hat{\Gamma}$.

Perhaps the simplest example of an abelian monoid is the additive monoid $\mathbb{Z}^+$ of non-negative integers. By definition a <u>free</u> abelian monoid with one generator, say X, is an abelian monoid Γ with an element $X \in \Gamma$ such that $n \mapsto X^n$ is an isomorphism of $\mathbb{Z}^+$ with Γ (where of course $X^0 = 1$). If for $P \in K\{\Gamma\}$ we write $P(X^n) = a_n$ so $P = \sum_n a_n X^n$ we see directly that $K\{\Gamma\} = K[X]$. For each $\alpha \in K$ there is clearly a unique $h \in \hat{\Gamma}$ such that $h(X) = \alpha$, namely, $X^n \mapsto \alpha^n$, so we see that $K[X]^\wedge = K\{\hat{\Gamma}\} = \hat{\Gamma} = K$ as before.

Next let Γ be an abelian monoid with elements $X_1, \ldots, X_n$ such that the map $\alpha \mapsto X^\alpha = X_1^{\alpha_1} \ldots X_n^{\alpha_n}$ is an isomorphism of $(\mathbb{Z}^+)^n$ with Γ. In this case we say Γ is a free abelian monoid on the generators $X_1, \ldots, X_n$ and clearly $K\{\Gamma\} = K[X_1, \ldots, X_n]$, the algebra of polynomials in n-variables. Given $\lambda = (\lambda_1, \ldots, \lambda_n) \in K^n$ the map $h_\lambda : \Gamma \to K$ defined by $X^\alpha \mapsto \lambda^\alpha = \lambda_1^{\alpha_1}\lambda_2^{\alpha_2}\ldots\lambda_n^{\alpha_n}$ is clearly in $\hat{\Gamma}$. Conversely, if $h \in \hat{\Gamma}$ and $h(X_i) = \lambda_i$ then $h = h_\lambda$ so $h \mapsto (h(X_1), \ldots, h(X_n))$ is a bijective map of $\hat{\Gamma}$ with K^n. Thus $K[X_1, \ldots, X_n]^\wedge = K\{\Gamma\}^\wedge = \hat{\Gamma} = K^n$.

We shall now carry this process of generalization to its logical conclusion and discuss the algebra of polynomials in a completely arbitrary set of variables. Because of their universal properties discussed below (every K-algebra is a homomorphic image of such an algebra) these algebras play an important role in what follows and will be referred to frequently.

Let J be an "index" set. If S is any set $\prod_{j \in J} S$ denotes the product of copies of S, one for each $j \in J$; i.e., the set of all functions $\lambda : J \to S$. We shall frequently use the notation $\{X_j\}_{j \in J}$ for an element of $\prod_{j \in J} S$, in which case we speak of an "indexed collection of elements of S (indexed by J)". By a <u>multi-index based on</u> J we mean an element α of $\prod_{j \in J} \mathbb{Z}^+$ such that $\alpha(j) = 0$ except for finitely many $j \in J$. We note that $\prod_{j \in J} \mathbb{Z}^+$ is an abelian monoid under addition and that the set of multi-indices based on J is a submonoid which we shall denote by $\coprod_{j \in J} \mathbb{Z}^+$.

If Γ is any abelian monoid, $\lambda \in \prod_{j \in J} \Gamma$, and α is a multi-index based on J we define an element $\lambda^\alpha \in \Gamma$ by $\lambda^\alpha = \prod_{\alpha(j) \neq 0} \lambda(j)^{\alpha(j)}$ (which is well defined because Γ is abelian). Elements of the form λ^α are called <u>monomials</u> in λ. We remark that by definition $\lambda^0 = 1$. The map $\alpha \to \lambda^\alpha$ is clearly a homomorphism of the (additive) abelian monoid $\coprod_{j \in J} \mathbb{Z}^+$ to the (multiplicative) abelian monoid Γ, and its image is called the sub-monoid of Γ generated by $\{\lambda(j)\}_{j\in J}$. In particular, if this monoid is Γ itself we say that $\{\lambda(j)\}_{j\in J}$ are generators for Γ (as a monoid) or generate Γ (as a monoid). [Caution: If $\mathcal{Q}$ is a K-algebra and $\lambda \in \prod_{j\in J} \mathcal{Q}$ then we say $\{\lambda(j)\}_{j\in J}$ generates $\mathcal{Q}$ as an <u>algebra</u> under the weaker condition that $\mathcal{Q}$ is all finite linear combinations of monomials in λ. Note that in either case $\{\lambda_j\}_{j\in J}$ generates C as a monoid (algebra) if and only if the smallest sub-monoid (sub-algebra) of C containing all the $\lambda(j)$ is C itself.] Clearly, if $\{\lambda(j)\}_{j\in J}$ generate Γ as a monoid they generate $K\{\Gamma\}$ as a K-algebra. Note that if Γ' is another abelian monoid and $h : \Gamma \to \Gamma'$ is a morphism of monoids (i.e., a multiplicative map such that $h(1) = 1$) then for any multi-index α, $h(\lambda^\alpha) = \sigma^\alpha$, where $\sigma = h \circ \lambda \in \prod_{j\in J} \Gamma'$. It follows of course that h is uniquely determined on the submodule generated by $\{\lambda(j)\}_{j\in J}$ by its values at the $\lambda(j)$.

An indexed set $\{X_j\}_{j\in J}$ of elements of Γ is called <u>free</u> in Γ if $X^\alpha \neq X^\beta$ whenever α and β are distinct multi-indices, i.e., if the morphism $\alpha \mapsto X^\alpha$ of $\coprod_{j\in J} \mathbb{Z}^+$ onto the submonoid of Γ generated by the $\{X_j\}_{j\in J}$ is an isomorphism (note that this implies that the X_j are distinct). If in addition $\{X_j\}_{j\in J}$ generates Γ then we say that Γ is the <u>free abelian monoid on generators</u> $\{X_j\}_{j\in J}$. Note that in particular this says that Γ is isomorphic to $\coprod_{j\in J} \mathbb{Z}^+$ (but written multiplicatively rather than additively).

Such a monoid Γ has an obvious but important characteristic property, namely, if Γ' is any other abelian monoid, then the set $\mathrm{Hom}(\Gamma, \Gamma')$ of module morphisms $\Gamma \to \Gamma'$ is in natural bijective correspondence with the set $\prod_{j\in J}\Gamma'$. Namely, given $h \in \mathrm{Hom}(\Gamma,\Gamma')$, $j \mapsto h(X_j)$ is an element of $\prod_{j\in J}\Gamma'$ which as pointed out above uniquely determines h since the X_j generate Γ. Conversely, given any $\lambda \in \prod_{j\in J}\Gamma'$ we can define a map $h:\Gamma\to\Gamma'$ as follows: if $\gamma\in\Gamma$ then $\gamma = X^{\alpha}$ for a unique multi-index α and we define $h(\gamma) = \lambda^{\alpha}$, so that h is given by $X^{\alpha}\mapsto\lambda^{\alpha}$ which is clearly in $\mathrm{Hom}(\Gamma,\Gamma')$. Moreover, we obviously have $h(X_j) = \lambda(j)$.

If we take for Γ' the multiplicative semi-group of the field K, then $\mathrm{Hom}(\Gamma, K) = \hat{\Gamma}$ the "dual" monoid of K-characters of Γ, thus for free abelian monoid on generators $\{X_j\}_{j\in J}$, $\hat{\Gamma} = \prod_{j\in J} K$ (where the " = " means is in natural bijective correspondence with).

Next consider the semi-group algebra $K\{\Gamma\}$ when Γ is freely generated by $\{X_j\}_{j\in J}$. Clearly, the general element of this algebra, which we shall also denote by $K[\{X_j\}_{j\in J}]$, can be written uniquely as a finite linear combination of monomials in the $\{X_j\}_{j\in J}$; i.e., it has the form $\sum_{\alpha} a_{\alpha}X^{\alpha}$ where the sum is over all multi-indices α based on J, the a_{α} are uniquely determined elements of K, and $a_{\alpha} = 0$ except for finitely many α. Given a second such element $\sum_{\alpha} b_{\alpha}X^{\alpha}$ their sum is of course $\sum_{\alpha}(a_{\alpha}+b_{\alpha})X^{\alpha}$ and their product is $\sum_{\alpha,\beta} a_{\alpha}b_{\beta}X^{\alpha+\beta} = \sum_{\gamma}(\sum_{\alpha+\beta=\gamma} a_{\alpha}b_{\beta})X^{\gamma}$. Clearly, what we have here is the algebra of all polynomials in the $\{X_j\}_{j\in J}$, which is how we shall henceforth refer to $K[\{X_j\}_{j\in J}]$.

Next recall the universal property of the semi-group algebra $K\{\Gamma\}$; namely, for any K-algebra $\mathcal{A}$ we have $\mathrm{Hom}(K\{\Gamma\},\mathcal{A}) = \mathrm{Hom}(\Gamma,\mathcal{A})$, where on the left "Hom" means K-algebra morphisms and on the right "Hom" means morphisms of abelian monoids and $\mathcal{A}$ is regarded as an abelian monoid under multiplication. When Γ is freely generated by $\{X_j\}_{j\in J}$ this gives:

$$\mathrm{Hom}(K[\{X_j\}_{j\in J}], \mathcal{A}) = \prod_{j\in J} \mathcal{A}$$

and in particular taking $\mathcal{A} = K$ gives

$$K[\{X_j\}_{j\in J}]^{\wedge} = \prod_{j\in J} K.$$

We restate this for convenience as a proposition.

1.2.13. <u>Proposition.</u> If $\mathcal{A}$ is an arbitrary K-algebra, then the set of algebra homomorphisms h of the polynomial algebra $K[\{X_j\}_{j\in J}]$ into $\mathcal{A}$ is in canonical bijective correspondence with the set $\prod_{j\in J} \mathcal{A}$ of maps λ of J into $\mathcal{A}$; namely, given h, $\lambda(j) = h(X_j)$ and given λ, $h(\Sigma_\alpha a_\alpha X^\alpha) = \Sigma_\alpha a_\alpha \lambda^\alpha$. In particular, (taking $\mathcal{A} = K$) the dual object $K[\{X_j\}_{j\in J}]^{\wedge}$ can be identified with $\prod_{j\in J} K$.

<u>Proof.</u> See 1.2.12. ■

We now have an analogue of 1.2.11.

1.2.14. <u>Proposition.</u> Identify $K[\{X_j\}_{j\in J}]^{\wedge}$ with $\prod_{j\in J} K$ as in 1.2.13. Then the Gelfand representation of $K[\{X_j\}_{j\in J}]$ is the "usual" representation which associates to the formal polynomial $\Sigma_\alpha a_\alpha X^\alpha$ the polynomial function $\prod_{j\in J} K \to K$ defined by $\lambda \mapsto \Sigma_\alpha a_\alpha \lambda^\alpha$. The set of all such polynomial functions (i.e., the image of the Gelfand representation) is denoted by $\mathcal{P}(\prod_{j\in J} K)$ and called the algebra of polynomial functions on $\prod_{j\in J} K$. A necessary and sufficient condition for $K[\{X_j\}_{j\in J}]$ to be strictly semi-simple is that K be infinite.

Proof. Only the final statement needs proof. If K is finite, say, $K = \{\alpha_1, \ldots, \alpha_n\}$ then let $X = X_j$ for any $j \in J$ and put $Q = (X-\alpha_1)(X-\alpha_2)\ldots(X-\alpha_n) \in K[\{X_j\}_{j\in J}]$. Then clearly Q is not zero but vanishes at every point of $\prod_{j\in J} K$. The fact that if K is infinite then the polynomial algebra is strictly semi-simple follows from the following more general result. ■

1.2.15. Lemma. If S is any infinite subset of K and $\sum_\alpha a_\alpha X^\alpha \in K[\{X_j\}_{j\in J}]$ is not the zero polynomial, then there exists $\lambda \in \prod_{j\in J} S$ such that $\sum_\alpha a_\alpha \lambda^\alpha \neq 0$.

Proof. Let us call the variables actually occurring in the polynomial $X_1, \ldots, X_n$ and proceed by induction on n. If $n = 0$ the polynomial is a non-zero constant and the result is trivial. Otherwise we can write the polynomial in the form $a_0(X_1, \ldots, X_{n-1}) + \ldots + a_k(X_1, \ldots, X_{n-1})X_n^k$ where $k \geq 1$ and the a_k are polynomials in $X_1, \ldots, X_{n-1}$ with $a_k \neq 0$. By induction we can find $\lambda_1, \ldots, \lambda_{n-1} \in S$ such that $a_k(\lambda_1, \ldots, \lambda_{n-1}) \neq 0$. Then we need only choose λ_n to be any element of S which is not one of the at most k roots of

$$a_0(\lambda_1, \ldots, \lambda_{n-1}) + \ldots + a_k(\lambda_1, \ldots, \lambda_{n-1})X_n^k.$$ ■

1.2.16. Remark. Let $\mathcal{A}$ be any K-algebra and let $\{\xi_j\}_{j\in J}$ be any set of generators for $\mathcal{A}$ as a K-algebra. If we form the polynomial algebra $K[\{X_j\}_{j\in J}]$ then by 1.2.13 there is a unique homomorphism $h^\xi : K[\{X_j\}_{j\in J}] \to \mathcal{A}$ such that $h^\xi(X_j) = \xi_j$, namely, $\Sigma a_\alpha X^\alpha \mapsto \Sigma a_\alpha \xi^\alpha$. Moreover, since the ξ_j generate $\mathcal{A}$ as an algebra, h^ξ is surjective. It follows (as pointed out in 1.0) that

$$(h^\xi)^\wedge : \hat{\mathcal{A}} \to K[\{X_j\}_{j\in J}]^\wedge = \prod_{j\in J} K$$

is injective. If we regard the ξ_j as functions on $\hat{\mathcal{A}}$ (via the Gelfand representation) then by 1.1.16 we have

$$(h^\xi)^\wedge(\varphi) = \{\hat{\xi}_j(\varphi)\}_{j\in J} = \{\varphi(\xi_j)\}_{j\in J}.$$

In other words the $\hat{\xi}_j$ are just the coordinates of the embedding of $\hat{\mathcal{A}}$ in $\prod_{j\in J} K$. This gives an important geometrical realization of $\hat{\mathcal{A}}$ as a subset of the generalized "affine space" $\prod_{j\in J} K$. We shall see later when we topologize $\hat{\mathcal{A}}$ (with the so-called Zariski topology) then this is in fact a topological embedding.

1.3. Z-closed Sets and Strict Radical Ideals.

If $\mathcal{A}$ is an algebra of K valued functions on some set S (i.e., $\mathcal{A}$ is a subalgebra of K^S) then there is an important correspondence between subsets of $\mathcal{A}$ and subsets of S and another going the other way. Namely, to each $T \subseteq \mathcal{A}$ we can associate the largest subset of S on which all the functions in T vanish (i.e., the intersection of all the "zero sets" $t^{-1}(0)$ for all $t \in T$), and to each $E \subseteq S$ we can associate the set of all functions in $\mathcal{A}$ which vanish identically on E. If $\mathcal{A}$ is not a priori an algebra of functions but is strictly semi-simple, then we can regard $\mathcal{A}$ via the Gelfand representation as an algebra of functions on $\hat{\mathcal{A}}$, and even if $\mathcal{A}$ is not strictly semi-simple we can regard $\mathcal{A}/\mathrm{Rad}_{\mathcal{A}}(0)$ as an algebra of functions on $\hat{\mathcal{A}}$. Below we examine some of the properties of these correspondences in the generality suggested.

1.3.1. Definition. Given a subset T of the algebra $\mathcal{A}$ we define a subset $V_K(T)$ of $\hat{\mathcal{A}}$ (or simply $V(T)$ when no confusion is likely) by:

$$\begin{aligned} V(T) &= \cap \{\hat{t}^{-1}(0) \mid t \in T\} \\ &= \{\varphi \in \hat{\mathcal{A}} \mid \hat{t}(\varphi) = 0 \quad \text{for all } t \in T\} \\ &= \{\varphi \in \hat{\mathcal{A}} \mid \varphi(t) = 0 \quad \text{for all } t \in T\} \\ &= \{\varphi \in \hat{\mathcal{A}} \mid \varphi \mid T = 0\} \\ &= \{\varphi \in \hat{\mathcal{A}} \mid T \subseteq \ker(\varphi)\}. \end{aligned}$$

Subsets of $\hat{\mathcal{A}}$ of the form V(T) will be called Z-closed.

1.3.2. Remark. We shall see below that the Z-closed subsets of $\hat{\mathcal{Q}}$ are in fact the closed sets of a topology for $\hat{\mathcal{Q}}$, the Zariski topology or simply the Z-topology for short.

1.3.3. Remark. Under the canonical bijection $\varphi \mapsto \ker(\varphi)$ of $\hat{\mathcal{Q}}$ with $\mathrm{Spec}(\mathcal{Q})$ (cf. 1.2.3) the subset $V(T)$ will correspond to a subset $\mathcal{V}(T)$ of $\mathrm{Spec}(\mathcal{Q})$. Clearly

$$\mathcal{V}(T) = \{M \in \mathrm{Spec}(\mathcal{Q}) \mid T \subseteq M\}.$$

1.3.4. Remark. It is immediate from the definition that if $T \subseteq \mathcal{Q}$ and $\mathcal{I}$ is the ideal of $\mathcal{Q}$ generated by T, then $V(T) = V(\mathcal{I})$. Thus to get all Z-closed sets we need only look at sets of the form $V(\mathcal{I})$ where $\mathcal{I}$ is an ideal. What we shall see below is that we do not even have to let $\mathcal{I}$ be an arbitrary ideal; it is enough to restrict attention to ideals which are intersections of strictly maximal ideals (so-called strict radical ideals). In fact it turns out that $\mathcal{I} \mapsto V(\mathcal{I})$ is a bijective correspondence between the Z-closed subsets of $\hat{\mathcal{Q}}$ and the strict radical ideals of $\mathcal{Q}$.

1.3.5. Definition. Given a subset S of $\hat{\mathcal{Q}}$ we define an ideal $I(S)$ of $\mathcal{Q}$ by:

$$
\begin{aligned}
I(S) &= \{x \in \mathcal{A} \mid (\hat{x} \mid S) = 0\} \\
&= \{x \in \mathcal{A} \mid \hat{x}(\varphi) = 0 \quad \text{all} \quad \varphi \in S\} \\
&= \{x \in \mathcal{A} \mid \varphi(x) = 0 \quad \text{all} \quad \varphi \in S\} \\
&= \bigcap \{\ker \varphi \mid \varphi \in S\}.
\end{aligned}
$$

Ideals of $\mathcal{A}$ of the form $I(S)$ will be called <u>strict radical ideals</u> of $\mathcal{A}$.

1.3.6. <u>Remark.</u> Under the canonical bijection $\varphi \to \ker(\varphi)$ of $\hat{\mathcal{A}}$ with $\text{Spec}(\mathcal{A})$ (cf. 1.2.3) the subsets S of $\hat{\mathcal{A}}$ will correspond bijectively to subsets $\mathcal{S}$ of $\text{Spec}(\mathcal{A})$. If we associate to $\mathcal{S}$ the ideal $\mathcal{I}(\mathcal{S}) = I(S)$ then clearly

$$\mathcal{I}(\mathcal{S}) = \bigcap \{M \mid M \in \mathcal{S}\}.$$

1.3.7. <u>Proposition.</u>

1) If $T_1 \subseteq T_2 \subseteq \mathcal{A}$ then $V(T_2) \subseteq V(T_1)$.
2) If $S_1 \subseteq S_2 \subseteq \hat{\mathcal{A}}$ then $I(S_2) \subseteq I(S_1)$.
3) If $T \subseteq \mathcal{A}$ then $T \subseteq I(V(T))$.
4) If $S \subseteq \hat{\mathcal{A}}$ then $S \subseteq V(I(S))$.

<u>Proof.</u> 1) says that the larger a set of functions, the smaller the set of points where they all vanish.

2) says that the larger a set of points, the smaller the set of functions which vanish identically on it.

3) says that the set of all functions $\hat{x}$ $(x \in \mathcal{A})$ which vanish at all points where all the functions $\hat{t}$ $(t \in T)$ vanish include in particular the functions $\hat{t}$.

4) says that the set of points φ at which vanish all the functions $\hat{x}$ which vanish identically on S includes in particular the points of S. $\blacksquare$

1.3.8. Remark. Let P and Q be two partially ordered sets and let $\varphi : P \to Q$ and $\psi : Q \to P$ be order reversing maps. Let us suppose also that $\psi(\varphi(p)) \geq p$ and $\varphi(\psi(q)) \geq q$ for all $p \in P$ and $q \in Q$. This pattern presents itself over and over again in mathematics and is called a "Galois Connection". (Consider the following case. P is the lattice of subfields of some field F which include some given ground field K. Let G be the group of automorphisms of F leaving K pointwise fixed, the so-called Galois group of F over K. Let Q be the lattice of subgroups of G. Given $p \in P$ let $\varphi(p)$ be the set of $g \in G$ leaving p pointwise fixed. Given $q \in Q$ let $\psi(q)$ be the set of $x \in F$ such that $gx = x$ for all $g \in q$.) There is only one real theorem concerning Galois connections and it is the following.

Theorem. Let $\varphi: P \to Q$ and $\psi : Q \to P$ be as above. Then $\mathrm{im}(\varphi)$ is the fixed point set of $\varphi \circ \psi : Q \to Q$ and similarly $\mathrm{im}(\psi)$ is the fixed point set of $\psi \circ \varphi : P \to P$. Moreover φ restricts to a bijective correspondence of $\mathrm{im}(\psi)$ with $\mathrm{im}(\varphi)$, whose inverse is the restriction of ψ.

Proof. Given $p \in P$ we have by assumption that $\psi(\varphi(p)) \geq p$. Since φ is order reversing $\varphi(\psi(\varphi(p)) \leq \varphi(p)$, or equivalently $\varphi \circ \psi(\varphi(p)) \leq \varphi(p)$. On the other hand for any $q \in Q$ we have $\varphi(\psi(q)) \geq q$ and in particular putting $q = \varphi(p)$ gives $\varphi \circ \psi(\varphi(p)) \geq \varphi(p)$. Together these show that $\mathrm{im}(\varphi)$ is included in the fixed point set of $\varphi \circ \psi$ and the reverse inclusion is trivial. Thus $\mathrm{im}(\varphi)$ has been identified with the fixed point set of $\varphi \circ \psi$ and by symmetry $\mathrm{im}(\psi)$ is the fixed point set of $\psi \circ \varphi$. Then if $p \in \mathrm{im}(\psi)$, $\psi \circ \varphi(p) = p$ while if $q \in \mathrm{im}(\varphi)$ then $\varphi \circ \psi(q) = q$. Together these show that $\varphi|\mathrm{im}(\psi)$ is a bijection $\mathrm{im}(\psi) \to \mathrm{im}(\varphi)$ and that $\psi|\mathrm{im}(\varphi)$ is its inverse. ∎

1.3.9. Theorem.

1) A subset $\mathcal{J}$ of $\mathcal{Q}$ is a strict radical ideal if and only if $\mathcal{J} = I(V(\mathcal{J}))$.

2) A subset S of $\hat{\mathcal{Q}}$ is Z-closed if and only if $S = V(I(S))$.

3) The maps $S \mapsto I(S)$ and $\mathcal{J} \mapsto V(\mathcal{J})$ are mutually inverse, inclusion reversing, bijective correspondences between the collection of Z-closed subsets of $\hat{\mathcal{Q}}$ and the set of strict radical ideals of $\mathcal{Q}$.

Proof. Letting P denote the lattice of subsets of $\mathcal{Q}$ and Q the lattice of subsets of $\hat{\mathcal{Q}}$ (both under inclusion of course) the content of 1.3.7 is that $V : P \to Q$ and $I : Q \to P$ define a Galois connection as defined in 1.3.8. Since by definition im(V) is the collection of Z-closed subsets of $\hat{\mathcal{Q}}$ and im(I) is the set of strict radical ideals of $\mathcal{Q}$, this theorem is a special case of the theorem proved in 1.3.8. ∎

1.3.10. Remark. In case the introduction of the operation V in 1.3.1 seemed unnatural or unmotivated, the next paragraph gives a plausible reason for introducing it. Recall from Section 1.0 that if $h : \mathcal{Q}_1 \to \mathcal{Q}_2$ is a surjective algebra homomorphism we have an induced injection of dual objects, $\hat{h} : \hat{\mathcal{Q}}_2 \to \hat{\mathcal{Q}}_1$. Now up to canonical identifications we can regard $\mathcal{Q}_2$ as $\mathcal{Q}_1/\mathcal{J}$ where $\mathcal{J} = \ker(h)$, and regard h as the canonical projection $\pi : \mathcal{Q}_1 \to \mathcal{Q}_1/\mathcal{J}$. If we normalize this way then we can describe $(\mathcal{Q}_1/\mathcal{J})^\wedge$ and $\hat{h}$ very explicitly.

1.3.11. Proposition. Let $\mathcal{J}$ be an ideal of $\mathcal{Q}$ and let $\pi : \mathcal{Q} \to \mathcal{Q}/\mathcal{J}$

be the canonical projection. Then

1) $(\mathcal{A}/\mathcal{I})^\wedge = \{\varphi \circ \pi^{-1} \mid \varphi \in V(\mathcal{I})\}$ and $\hat{\pi} : (\mathcal{A}/\mathcal{I})^\wedge \to \hat{\mathcal{A}}$ is the injective map $\varphi \circ \pi^{-1} \mapsto \varphi$. In particular $\hat{\pi}$ maps $(\mathcal{A}/\mathcal{I})^\wedge$ one-to-one onto' $V(\mathcal{I})$, thereby giving a canonical identification of $(\mathcal{A}/\mathcal{I})^\wedge$ with $V(\mathcal{I})$.

2) If we identify $(\mathcal{A}/\mathcal{I})^\wedge$ with $V(\mathcal{I})$ as above then the Gelfand representation of $\mathcal{A}/\mathcal{I}$ is given explicitly in terms of the Gelfand representation of $\mathcal{A}$ by the formula $\widehat{\pi(x)} = \hat{x} \mid V(\mathcal{I})$.

Proof. Recalling from Section 1.0 that $\hat{\pi}(\varphi) = \varphi \circ \pi$ and that by definition $V(\mathcal{I}) = \{\varphi \in \hat{\mathcal{A}} \mid \mathcal{I} \subseteq \ker \varphi\}$, the first statement is a triviality and the second follows from 1.1.16. ∎

1.3.12. Definition. To each ideal $\mathcal{I}$ of $\mathcal{A}$ we associate an ideal $\text{Rad}_{\mathcal{A}}(\mathcal{I})$ including $\mathcal{I}$, called the strict radical of $\mathcal{I}$ in $\mathcal{A}$, by

$$\text{Rad}_{\mathcal{A}}(\mathcal{I}) = \ker(\mathcal{A} \to \mathcal{A}/\mathcal{I} \to K^{(\mathcal{A}/\mathcal{I})^\wedge})$$

where $\mathcal{A} \to \mathcal{A}/\mathcal{I}$ is canonical and $\mathcal{A}/\mathcal{I} \to K^{(\mathcal{A}/\mathcal{I})^\wedge}$ is the Gelfand representation of $\mathcal{A}/\mathcal{I}$.

1.3.13. Proposition. If $\mathcal{I}$ is an ideal of $\mathcal{A}$ then

$$\text{Rad}_{\mathcal{A}}(\mathcal{I}) = I(V(\mathcal{I})) = \mathcal{I}(V(\mathcal{I})).$$

In other words, $\text{Rad}_{\mathcal{A}}(\mathcal{I})$ is the intersection of all the strictly maximal ideals of $\mathcal{A}$ which include $\mathcal{I}$.

Proof. By 1.3.11 the homomorphism $\mathcal{A} \to \mathcal{A}/\mathcal{I} \to K^{(\mathcal{A}/\mathcal{I})^\wedge}$ is given essentially by $x \mapsto \hat{x} | V(\mathcal{I})$. Thus

$$\begin{aligned} x \in \mathrm{Rad}_{\mathcal{A}}(\mathcal{I}) &\iff \hat{x}(\varphi) = 0 \quad \text{all} \quad \varphi \in V(\mathcal{I}) \\ &\iff \varphi(x) = 0 \quad \text{all} \quad \varphi \in V(\mathcal{I}) \\ &\iff x \in \bigcap \{\ker \varphi \mid \varphi \in V(\mathcal{I})\} \\ &\iff x \in I(V(\mathcal{I})). \end{aligned}$$

■

1.3.14. Remark. Note that 1.3.13 justifies calling $\mathrm{Rad}_{\mathcal{A}}(\mathcal{I})$ the strict radical of $\mathcal{I}$. In fact by 1.3.9 $\mathrm{Rad}_{\mathcal{A}}(\mathcal{I})$ is a strict radical ideal of $\mathcal{A}$ and conversely every strict radical ideal of $\mathcal{A}$ is of the form $\mathrm{Rad}_{\mathcal{A}}(\mathcal{I})$. In fact $\mathcal{I}$ is a strict radical ideal of $\mathcal{A}$ if and only if $\mathcal{I} = \mathrm{Rad}_{\mathcal{A}}(\mathcal{I})$.

1.3.15. Proposition. If $\mathcal{I}$ is an ideal of $\mathcal{A}$ then $\mathcal{I} \subseteq \mathrm{Rad}_{\mathcal{A}}(\mathcal{I})$ and

$$\mathrm{Rad}_{\mathcal{A}/\mathcal{I}}(0) = \mathrm{Rad}_{\mathcal{A}}(\mathcal{I})/\mathcal{I}.$$

Thus $\mathcal{A}/\mathcal{I}$ is strictly semi-simple if and only if $\mathcal{I} = \mathrm{Rad}_{\mathcal{A}}(\mathcal{I})$, i.e., if and only if $\mathcal{I}$ is a strict radical ideal of $\mathcal{A}$.

Proof. Immediate from the definition of $\mathrm{Rad}_{\mathcal{A}}(\mathcal{I})$, 1.3.9, and 1.3.13. ■

1.3.16. Proposition. If $\mathcal{I}$ and $\mathcal{I}'$ are ideals of $\mathcal{A}$ with $\mathcal{I} \subseteq \mathcal{I}'$ then $\mathrm{Rad}_{\mathcal{A}}(\mathcal{I}) \subseteq \mathrm{Rad}_{\mathcal{A}}(\mathcal{I}')$. Moreover $\mathrm{Rad}_{\mathcal{A}}(\mathcal{I}) = \mathrm{Rad}_{\mathcal{A}}(\mathcal{I}')$ if and only if $\mathcal{I}' \subseteq \mathrm{Rad}_{\mathcal{A}}(\mathcal{I})$. Thus $\mathrm{Rad}_{\mathcal{A}}(\mathcal{I})$ is the smallest ideal of $\mathcal{A}$ which includes $\mathcal{I}$ and whose quotient is strictly semi-simple.

Proof. Immediate from the fact that $\mathrm{Rad}_{\mathcal{A}}(\mathcal{I}) = \bigcap \{M \in \mathrm{Spec}(\mathcal{A}) \mid \mathcal{I} \subseteq M\}$. ■

1.3.17. Theorem. Let $\mathcal{A}$ be strictly semi-simple and have only finitely many different minimal prime ideals $p_1, p_2, \ldots, p_s$. Then each p_i is a strict radical ideal of $\mathcal{A}$.

Proof. Let $q_i = I(V(p_i)) = \bigcap \{M \in \mathrm{Spec}(\mathcal{A}) \mid p_i \subseteq M\}$, so that $p_i \subseteq q_i$ and what we must show (cf., 1.3.9) is the reverse inclusion $q_i \subseteq p_i$. It will even suffice to show that for some $j = 1, 2, \ldots, s$ we have $q_j \subseteq p_i$. For since $p_j \subseteq q_j \subseteq p_i$ and the p_k are distinct minimal prime ideals it will follow that $p_j = p_i$ and so $i = j$. Now note that every $M \in \mathrm{Spec}(\mathcal{A})$ includes at least one of the p_k (for M itself is maximal, hence prime, so by Zorn's lemma there is a maximal chain of prime ideals included in M. The intersection of this, or any, chain of prime ideals is easily seen to be prime, and by maximality of the chain it is even a minimal prime of $\mathcal{A}$, i.e., one of the p_k). It follows that

$$\bigcap_{k=1}^{s} q_k = \bigcap_{k=1}^{s} \bigcap \{M \in \mathrm{Spec}(\mathcal{A}) \mid p_k \subseteq M\}$$

$$= \bigcap \{M \mid M \in \mathrm{Spec}(\mathcal{A})\}$$

$$= \mathrm{Rad}_{\mathcal{A}}(0).$$

But by assumption $\mathcal{A}$ is strictly semi-simple, i.e., $\mathrm{Rad}_{\mathcal{A}}(0) = 0$, so $\bigcap_{k=1}^{s} q_k = 0$ and a fortiori $q_1 q_2 \cdots q_s \subseteq 0 \subseteq p_i$. But p_i is prime, so that if it includes a product of ideals it must include one of them; i.e., $q_j \subseteq p_i$ for some $j = 1, 2, \ldots, s$. ∎

1.4. The Z-topology.

1.4.1. Proposition. Let $\mathcal{I}_1$ and $\mathcal{I}_2$ be ideals of $\mathcal{A}$ and let $\mathcal{I} = \mathcal{I}_1\mathcal{I}_2$; i.e., $\mathcal{I}$ is the ideal of finite sums of products xy with $x \in \mathcal{I}_1$ and $y \in \mathcal{I}_2$. Then $V(\mathcal{I}) = V(\mathcal{I}_1) \cup V(\mathcal{I}_2)$. In particular the collection of Z-closed subsets of $\hat{\mathcal{A}}$ is stable under the taking of finite unions.

Proof. Given $x \in \mathcal{I}_1$ and $y \in \mathcal{I}_2$ we have $(xy)^\wedge = \hat{x}\hat{y}$ and since $\hat{x}$ vanishes on $V(\mathcal{I}_1)$ and $\hat{y}$ vanishes on $V(\mathcal{I}_2)$ it follows that $(xy)^\wedge$ vanishes on their union. Thus $V(\mathcal{I}_1) \cup V(\mathcal{I}_2) \subseteq V(\mathcal{I})$. Given $\varphi \notin V(\mathcal{I}_1) \cup V(\mathcal{I}_2)$ we can (since $\varphi \notin V(\mathcal{I}_1)$) find $x \in \mathcal{I}_1$ with $\hat{x}(\varphi) \neq 0$ and similarly we can find $y \in \mathcal{I}_2$ such that $\hat{y}(\varphi) \neq 0$. Then $(xy)^\wedge(\varphi) \neq 0$ so that, since $xy \in \mathcal{I}$, $\varphi \notin V(\mathcal{I})$. Thus we also have $V(\mathcal{I}) \subseteq V(\mathcal{I}_1) \cup V(\mathcal{I}_2)$. ∎

1.4.2. Proposition. Let $\{\mathcal{I}_\alpha\}$ be a family of ideals of $\mathcal{A}$ and let $\mathcal{I} = \Sigma_\alpha \mathcal{I}_\alpha$ be the ideal they generate, i.e., the ideal of all sums $\Sigma_\alpha x_\alpha$ with $x_\alpha \in \mathcal{I}_\alpha$ and $x_\alpha = 0$ except for finitely many α. Then $V(\mathcal{I}) = \bigcap_\alpha V(\mathcal{I}_\alpha)$. In particular the collection of Z-closed subsets of $\hat{\mathcal{A}}$ is stable under taking arbitrary intersections.

Proof. $(\Sigma_\alpha x_\alpha)^\wedge = \Sigma_\alpha \hat{x}_\alpha$ and since $\hat{x}_\alpha$ vanishes on $V(\mathcal{I}_\alpha)$ it follows that $(\Sigma_\alpha x_\alpha)^\wedge$ vanishes on $\bigcap_\alpha V(\mathcal{I}_\alpha)$. Thus $\bigcap_\alpha V(\mathcal{I}_\alpha) \subseteq V(\mathcal{I})$. If $\varphi \notin \bigcap_\alpha V(\mathcal{I}_\alpha)$ then choose α_0 with $\varphi \notin V(\mathcal{I}_{\alpha_0})$ and next pick $x_{\alpha_0} \in \mathcal{I}_{\alpha_0}$ such that $\hat{x}_{\alpha_0}(\varphi) \neq 0$. Then $x_{\alpha_0} \in \mathcal{I}$ so $\varphi \notin V(\mathcal{I})$. It follows that we also have $V(\mathcal{I}) \subseteq \bigcap_\alpha V(\mathcal{I}_\alpha)$. ∎

1.4.3. Proposition. $\hat{\mathcal{A}} = V((0))$ and $\phi = V(\mathcal{A})$. Thus $\hat{\mathcal{A}}$ and ϕ are Z-closed subsets of $\hat{\mathcal{A}}$.

Proof. Trivial. (Recall that for $\varphi \in \hat{\mathcal{A}}$, $\varphi(1) = 1 \neq 0$, or $\hat{1}(\varphi) \neq 0$ so $\varphi \notin V(\mathcal{A})$). ■

1.4.4. Proposition. If $\varphi \in \hat{\mathcal{A}}$ then $V(\ker(\varphi)) = \{\varphi\}$. Thus $\{\varphi\}$ is a Z-closed subset of $\hat{\mathcal{A}}$.

Proof. Recall from 1.3.1 that $\psi \in V(\ker(\varphi))$ if and only if $\psi | \ker(\varphi) = 0$, i.e., if and only if $\ker(\varphi) \subseteq \ker(\psi)$. Since $\ker(\varphi)$ is a maximal ideal of $\mathcal{A}$ this is the same as $\ker(\varphi) = \ker(\psi)$ which by 1.2.3 is the same as $\varphi = \psi$. ■

1.4.5. Theorem. The Z-closed subsets of $\hat{\mathcal{A}}$ are the closed sets for a T_1-topology, called the Zariski topology or simply the Z-topology for $\hat{\mathcal{A}}$. Subsets of $\hat{\mathcal{A}}$ of the form $\hat{\mathcal{A}}_x = \{\varphi \in \hat{\mathcal{A}} \,|\, \hat{x}(\varphi) \neq 0\}$ where $x \in \mathcal{A}$, are called basic open sets and form a base for the Z-topology.

Proof. It is immediate from 1.4.1, 1.4.2, 1.4.3 and 1.4.4 that the Z-closed sets are the closed sets of a T_1-topology. By 1.3.1 a set is Z-closed if and only if it is the intersection of a collection of complements of basic open sets, so a set is Z-open if and only if it is a union of basic open sets. ■

1.4.6. Proposition. Every set X has a weakest T_1-topology, and this topology is compact. Its closed sets are X itself and the finite subsets of X.

Proof. Trivial.

1.4.7. Example. The Z-topology for $K[X]^\wedge$ is its weakest T_1-topology. To see this identify $K[X]^\wedge$ with K via the map $\varphi \mapsto \varphi(X)$ (cf., 1.2.11) and recall that with this identification the image of $f(X) \in K[X]$ under the Gelfand representation is just the map $\alpha \mapsto f(\alpha)$ of K into K. Now Z-closed sets are just intersections of sets of the form $f^{-1}(0) = \{\alpha \in K \mid f(\alpha) = 0\}$ for $f \in K[X]$. But such sets are either all of K, if $f(X) = 0$, or else a finite set. Moreover, any finite subset $\{\alpha_1, \ldots, \alpha_k\}$ of K is of the form $f^{-1}(0)$ with $f(X) = \prod(X-\alpha_i)$.

1.4.8. Proposition. The Z-topology for $\hat{\mathcal{a}}$ is the weakest topology for $\hat{\mathcal{a}}$ making $\hat{x} : \hat{\mathcal{a}} \to K$ continuous for all $x \in \mathcal{a}$ when K is given its weakest T_1-topology.

Proof. To see that $\hat{x}$ is Z-continuous we must show that $\hat{x}^{-1}(\{\alpha_1, \ldots, \alpha_k\})$ is Z-closed. Since it is the union of the $\hat{x}^{-1}(\alpha_i) = ((x-\alpha_i)^\wedge)^{-1}(0)$, which is Z-closed by definition, this is clear. Conversely, the closed sets of any topology for $\hat{\mathcal{a}}$ making the $\hat{x}$ continuous must include the $\hat{x}^{-1}(0)$, which generate the closed sets of the Z-topology. ■

1.4.9. Remark. Suppose K is a topological field. We can then define the W-topology for $\hat{\mathcal{a}}$ as the weakest topology making each function $\hat{x} : \hat{\mathcal{a}} \to K$ continuous. By 1.4.8 the W-topology is always at least as strong as the Z-topology.

1.4.10. Remark. There is another way of looking at the Z-topology for $\hat{\mathcal{A}}$ that is interesting. As above let K have its weakest T_1-topology. Let $X = \prod_{x \in \mathcal{A}} K$ denote the product of copies of K, one for each $x \in \mathcal{A}$; explicitly, X is just the set $K^{\mathcal{A}}$ of functions $f : \mathcal{A} \to K$. For each $x \in \mathcal{A}$ we let $\Pi_x : X \to K$ denote the "projection" map $f \mapsto f(x)$. We give X its Tychonoff topology, i.e., the weakest topology making each of the maps Π_x continuous. Since K is compact it follows from Tychonoff's theorem that X is compact. Now note that $\hat{\mathcal{A}} \subseteq X$. Let us consider $\Pi_x | \hat{\mathcal{A}}$; it is given by $\varphi \mapsto \varphi(x) = \hat{x}(\varphi)$, i.e., $\Pi_x | \hat{\mathcal{A}} = \hat{x}$. It follows from 1.4.8 that $\hat{\mathcal{A}}$ with the Z-topology is actually a subspace of X. It is now tempting to conclude that $\hat{\mathcal{A}}$ is "obviously" closed in X, hence compact in the Z-topology. The argument goes as follows: $f \in X$ belongs to $\hat{\mathcal{A}}$ if and only if for all $x, y \in \mathcal{A}$ we have $f(x+y) = f(x)+f(y)$ and $f(xy) = f(x)f(y)$, in other words, if and only if f satisfies all the equations $\Pi_{x+y} = \Pi_x + \Pi_y$ and $\Pi_{xy} = \Pi_x \Pi_y$. The problem is that, whereas the individual maps Π_z are each continuous (by definition) it does not follow that $\Pi_x + \Pi_y$ and $\Pi_x \Pi_y$ are continuous unless K is a topological algebra. Now if K is finite, then K is discrete and hence a topological algebra. On the other hand, if K is infinite then K is T_1 but not Hausdorff, so in this case K is certainly not a topological algebra (recall that a T_1 topological group is even completely regular). Thus if K is finite $\hat{\mathcal{A}}$ will always be compact while if K is infinite we cannot conclude this is the case. In fact we shall now see that we can always find strictly semi-simple algebras $\mathcal{A}$ such that $\hat{\mathcal{A}}$ is not compact.

1.4.11. Example. Assume K is infinite and recall that, by 1.2.14, for any set J the polynomial algebra $K[\{X_j\}_{j\in J}]$ is strictly semi-simple and $K[\{X_j\}_{j\in J}]^\wedge$ is naturally isomorphic to $\prod_{j\in J} K$. That is $K[\{X_j\}_{j\in J}]$ is isomorphic to the algebra $\mathcal{P}(\prod_{j\in J} K)$ of "polynomial functions" on $\prod_{j\in J} K$. The isomorphism (the Gelfand representation) is the usual one which associates to each formal polynomial $\Sigma_\alpha a_\alpha X^\alpha$ the polynomial function $\prod_{j\in J} K \to K$ given by $\lambda \mapsto \Sigma_\alpha a_\alpha \lambda^\alpha$. Now take $J = K \cup \{\infty\}$ where ∞ is any element not in K. For each $\alpha \in K$ let F_α denote the Z-closed subset of $\prod_{j\in J} K$ defined by $X_\alpha(X_\infty - \alpha) = 1$, i.e., the set of $\xi = \{\xi_j\}_{j\in J}$ in $\prod_{j\in J} K$ such that $\xi_\infty \neq \alpha$ and $\xi_\alpha = 1/(\xi_\infty - \alpha)$. Given any finite set $\{\alpha_1, \ldots, \alpha_n\} \subseteq K \subseteq J$ we can always find α_0 in K which is different from any of $\alpha_1, \ldots, \alpha_n$ (because K is infinite). Then any $\xi \in \prod_{j\in J} K$ such that $\xi_\infty = \alpha_0$ and $\xi_{\alpha_i} = 1/(\alpha_i - \alpha_0)$ for $i = 1, \ldots, n$ is in the intersection of $F_{\alpha_1}, \ldots, F_{\alpha_n}$, so the F_{α_i} have the finite intersection property. However, no element ξ of $\prod_{j\in J} K$ can belong to all the F_α; in fact if $\xi_\infty = \alpha$ then ξ cannot belong to F_α since we have $\xi_\alpha(\xi_\infty - \alpha) = 0$, and not $\xi_\alpha(\xi_\infty - \alpha) = 1$. Thus $\prod_{j\in J} K = K[\{X_j\}_{j\in J}]^\wedge$ is not compact in the Z-topology.

1.4.12. Proposition. If $E \subseteq \hat{\mathcal{A}}$ then the closure of E in the Z-topology is the set $V(I(E))$ of $\varphi \in \hat{\mathcal{A}}$ such that if $x \in \mathcal{A}$ and $\hat{x}|E = 0$ then $\hat{x}(\varphi) = 0$.

Proof. We know $E \subseteq V(I(E))$ from 1.3.7 and of course $V(I(E))$ is Z-closed by definition. If F is Z-closed and $E \subseteq F$ then $I(F) \subseteq I(E)$ so $V(I(E)) \subseteq V(I(F))$, both by 1.3.7 again. But then by 1.3.9 $V(I(E)) \subseteq F$. This shows $V(I(E))$ is the smallest Z-closed set which includes E. ■

1.4.13. Corollary. A subset F of $\hat{\mathcal{A}}$ is Z-closed if and only if given $\varphi \in \hat{\mathcal{A}} - F$ there exists $x \in \mathcal{A}$ such that $\hat{x}(\varphi) \neq 0$ and $\hat{x} | F = 0$. A subset $\mathcal{O}$ of $\hat{\mathcal{A}}$ is Z-open if and only if given $\varphi \in \mathcal{O}$ there exists $x \in \mathcal{A}$ such that $\hat{x}(\varphi) \neq 0$ and $\hat{x} | (\hat{\mathcal{A}} - \mathcal{O}) = 0$, i.e., $\varphi \in \hat{\mathcal{A}}_x \subseteq \mathcal{O}$ where $\hat{\mathcal{A}}_x$ is the basic open set $\{\psi \in \hat{\mathcal{A}} \,|\, \hat{x}(\psi) \neq 0\}$.

We next consider how certain standard operations on algebras $\mathcal{A}$ are reflected in their dual objects $\hat{\mathcal{A}}$, considered as topological spaces with the Z-topology.

1.4.14. Proposition. If $h : \mathcal{A}_1 \to \mathcal{A}_2$ is a homomorphism of algebras then $\hat{h} : \hat{\mathcal{A}}_2 \to \hat{\mathcal{A}}_1$ is continuous in the Z-topologies. If h is surjective then $\hat{h}$ is a homeomorphism into. (In particular if $\mathcal{I}$ is an ideal in $\mathcal{A}$ then with the canonical identification of $(\mathcal{A}/\mathcal{I})^\wedge$ with $V(\mathcal{I}) \subseteq \hat{\mathcal{A}}$, the Z-topology for $\hat{\mathcal{A}}$ induces the Z-topology of $(\mathcal{A}/\mathcal{I})^\wedge$; i.e., $(\mathcal{A}/\mathcal{I})^\wedge$ is a topological subspace of $\hat{\mathcal{A}}$.) If h is injective and $\mathcal{A}_2$ is strictly semi-simple then $\operatorname{im}(\hat{h})$ is Z-dense in $\hat{\mathcal{A}}_1$.

Proof. If $x \in \mathcal{A}_1$ then $\hat{x} \circ \hat{h} = (h(x))^\wedge : \hat{\mathcal{A}}_2 \to K$ is continuous by 1.4.8 and so, also by 1.4.8, it follows that $\hat{h}$ is continuous. That $\hat{\Pi} : (\mathcal{A}/\mathcal{I})^\wedge \to \hat{\mathcal{A}}$ is a homeomorphism onto its image $V(\mathcal{I})$ with the induced Z-topology follows from 1.3.11 and 1.4.8. That $\hat{h}$ is a homeomorphism into when h is surjective is only a superficially more general statement. To show that $\text{im}(\hat{h})$ is dense in $\hat{\mathcal{A}}_1$ under the stated conditions it will suffice to show that given $x \in \mathcal{A}_1$ if $\hat{x}|\text{im}(\hat{h}) = 0$ then $\hat{x} = 0$. Now $\hat{x}|\text{im}(\hat{h}) = 0$ means $0 = \hat{x} \circ \hat{h} = (h(x))^\wedge$. Since $\mathcal{A}_2$ is strictly semi-simple $h(x) = 0$ and since h is injective $x = 0$, so $\hat{x} = 0$. ■

1.4.15. Remark. Given an arbitrary K-algebra $\mathcal{A}$ let $\{x_\alpha\}_{\alpha\in A}$ be a set of generators for $\mathcal{A}$ (i.e., a collection of elements of $\mathcal{A}$ such that every element of $\mathcal{A}$ can be expressed as a polynomial in the x_α, or equivalently such that no proper subalgebra of $\mathcal{A}$ contains all the x_α). There is then a unique homomorphism $h : K[\{X_\alpha\}_{\alpha\in A}]$ onto $\mathcal{A}$ mapping X_α onto x_α. The kernel $\mathcal{J}$ of h is of course the set of all polynomial relations satisfied by the x_α. Identifying $K[\{X_\alpha\}_{\alpha\in A}]^\wedge$ with $\prod_{\alpha\in A} K$ (cf., 1.2.13) we have an injective map $\hat{h} : \hat{\mathcal{A}} \to \prod_{\alpha\in A} K$ which is actually a topological embedding when both are given their Z-topologies. Thus we can identify $\hat{\mathcal{A}}$ with the set of all $\{\xi_\alpha\}_{\alpha\in A}$ in $\prod_{\alpha\in A} K$ such that $P(\xi_\alpha) = 0$ for all $P \in \mathcal{J}$, (cf., 1.3.11). We consider next an illustrative class of examples.

Let $P(X, Y)$ be an irreducible polynomial in $K[X, Y]$. We assume $V = \{(\xi_1, \xi_2) \in K^2 \mid P(\xi_1, \xi_2) = 0\}$ is infinite (so of course K is infinite). Let $P(X, Y) = a_n(X)Y^n + \ldots + a_0(X)$. Note that $a_n(\xi) = \ldots = a_0(\xi) = 0$ for $\xi \in K$ is impossible (for then $(X-\xi)$ would divide $P(X, Y)$), hence for each ξ in K $P(\xi, Y) = a_n(\xi)Y^n + \ldots + a_0(\xi)$ has at most n roots and hence the set of $\xi \in K$ such that $P(\xi, Y)$ has at least one root is infinite. We shall see later (2.3.13) that P generates the ideal $\mathcal{I} = \mathcal{I}(V)$ of all $Q \in K[X, Y]$ which vanish on V. Thus $\mathcal{I}$ is a strict radical ideal in $K[X, Y]$ and so $\mathcal{A} = K[X, Y]/(P)$ is strictly semi-simple. By the above we have a natural identification of $\hat{\mathcal{A}}$ with V; $(\xi_1, \xi_2) \in V$ corresponds to the homomorphism $\varphi : \mathcal{A} \to K$ taking the coset $Q(X, Y) + (P)$ to $Q(\xi_1, \xi_2)$. Next note that since the subalgebra $K[X]$ of $K[X, Y]$ is disjoint from (P) it injects into $\mathcal{A}$. Let us call its image B; so B is the subalgebra of $\mathcal{A}$ consisting of elements of the form $f(X) + (P)$. We know that $\hat{B}$ can be canonically identified with K, an element $\xi \in K$ corresponding to the homomorphism $\psi : B \to K$ mapping $f(X) + (P) \mapsto f(\xi)$. Note that if $j : B \to \mathcal{A}$ is the inclusion map then with the above identifications ($\hat{\mathcal{A}} = V \subseteq K \times K$ and $\hat{B} = K$) the restriction map $\hat{j} : \hat{\mathcal{A}} \to \hat{B}$ ($\varphi \mapsto \psi = \varphi \circ j = \varphi | B$) is clearly $(\xi_1, \xi_2) \mapsto \xi_1$, i.e., projection on the first factor. Thus $\mathrm{im}(\hat{j})$ is the set of $\xi \in K$ such that $P(\xi, Y)$ has at least one root in K, which as we have seen is an infinite subset of K, and hence Z-dense in K as expected from 1.4.14. If $P(X, Y) = XY-1$ (so V is the usual hyperbola) then $\mathrm{im}(\hat{j}) = K-\{0\} = \hat{B}-\{\text{point}\}$. So we see that in general $\hat{j}$ is not surjective, no matter what K is. If K is algebraically closed then $\mathrm{im}(\hat{j})$ has a finite complement, namely, the set of points $\xi \in K$, $a_1(\xi) = a_2(\xi) = \ldots = a_n(\xi) = 0$. In particular if $a_n(X) = 1$ then $\hat{j}$ is surjective when K is algebraically closed. If $K = \mathbb{R}$ and $a_n(X) = 1$ then $\hat{j}$ is again surjective provided n is odd (a polynomial of odd degree over $\mathbb{R}$ always has a root). However, if n is even $\mathrm{im}(\hat{j})$ can be quite "small". For example if $P(X,Y) = X^2+Y^2-r^2$ $(r > 0)$ then $\mathrm{im}(\hat{j}) = [-r,r]$. If $P(X,Y) = Y^2-X$ then $\mathrm{im}(\hat{j}) = [0,\infty)$.

We shall come back frequently to the study of the nature of $\mathrm{im}(\hat{j})$, where $j : B \to \mathcal{A}$ is the inclusion map of a subalgebra B in a strictly semi-simple algebra $\mathcal{A}$. The above examples are provided to indicate that this is a non-trivial, somewhat subtle question.

1.4.16. Proposition. Let $\mathcal{A}_1$ and $\mathcal{A}_2$ be algebras and $\mathcal{A} = \mathcal{A}_1 \oplus \mathcal{A}_2$ their direct sum. Let Π_1 and Π_2 be the canonical projections of $\mathcal{A}$ on $\mathcal{A}_1$ and $\mathcal{A}_2$ respectively, and identify $\hat{\mathcal{A}}_i$ with its image in $\hat{\mathcal{A}}$ under $\hat{\Pi}_i$ (i.e., we identify $\varphi \in \hat{\mathcal{A}}_1$ with an element of $\hat{\mathcal{A}}$ by $\varphi(x,y) = \varphi(x)$ and similarly $\psi \in \hat{\mathcal{A}}_2$ is identified with an element of $\hat{\mathcal{A}}$ by $\psi(x,y) = \psi(y)$). Then $\hat{\mathcal{A}}$ with its Z-topology is the topological sum of $\hat{\mathcal{A}}_1$ and $\hat{\mathcal{A}}_2$ with their Z-topologies.

Proof. Note that $\hat{\mathcal{A}}_1$ is the set of $\varphi \in \hat{\mathcal{A}}$ which vanish on $\mathcal{A}_2$ and similarly $\hat{\mathcal{A}}_2$ is the set of $\varphi \in \mathcal{A}$ which vanish on $\mathcal{A}_1$. Thus an element of $\hat{\mathcal{A}}_1 \cap \hat{\mathcal{A}}_2$ would vanish on $\mathcal{A}_1$ and $\mathcal{A}_2$ and so be identically zero, which is impossible. By 1.4.14 it remains only to show that any element $\varphi \in \hat{\mathcal{A}}$ vanishes either on $\mathcal{A}_1$ or on $\mathcal{A}_2$. Now for any $x \in \mathcal{A}_1$ and $y \in \mathcal{A}_2$ we have $0 = \varphi(0,0) = \varphi((x,0)(0,y)) = \varphi(x,0)\varphi(0,y)$. If $\varphi \notin \hat{\mathcal{A}}_2$ then there exists $x \in \mathcal{A}_1$ such that $\varphi(x,0) \neq 0$ and it follows that $\varphi(0,y) = 0$ for all $y \in \mathcal{A}_2$, i.e., $\varphi \in \hat{\mathcal{A}}_1$. ∎

1.4.17. Remark. If $\mathcal{A}_1$ and $\mathcal{A}_2$ are two algebras recall that their tensor product $\mathcal{A}_1 \otimes \mathcal{A}_2$ (as vector spaces over K) has a unique structure of algebra such that $(x_1 \otimes y_1)(x_2 \otimes y_2) = x_1x_2 \otimes y_1y_2$. When we speak of the tensor product of algebras we will always mean this algebra. If B is any algebra and $\varphi : \mathcal{A}_1 \to B$ and $\psi : \mathcal{A}_2 \to B$ are homomorphisms, then there is a unique homomorphism $\varphi \otimes \psi : \mathcal{A}_1 \otimes \mathcal{A}_2 \to B$ such that $\varphi \otimes \psi(x \otimes y) = \varphi(x)\psi(y)$. Moreover, we have canonical injections $x \mapsto (x, 1)$ and $y \mapsto (1 \otimes y)$ of $\mathcal{A}_1$ and $\mathcal{A}_2$ into $\mathcal{A}_1 \otimes \mathcal{A}_2$, so conversely if $\lambda : \mathcal{A}_1 \otimes \mathcal{A}_2 \to B$ is a homomorphism then $x \mapsto \lambda(x \otimes 1)$ and $y \mapsto \lambda(1 \otimes y)$ are homomorphisms $\varphi : \mathcal{A}_1 \to B$ and $\psi : \mathcal{A}_2 \to B$ such that $\lambda = \varphi \otimes \psi$. In other words we have $\mathrm{Hom}(\mathcal{A}_1 \otimes \mathcal{A}_2, B) = \mathrm{Hom}(\mathcal{A}_1, B) \times \mathrm{Hom}(\mathcal{A}_2, B)$. In particular taking $B = K$ we get the following result.

1.4.18. <u>Proposition.</u> Let $\mathcal{A}_1$ and $\mathcal{A}_2$ be two algebras. Given $\lambda \in (\mathcal{A}_1 \otimes \mathcal{A}_2)^\wedge$ define $\lambda_1 \in \hat{\mathcal{A}}_1$ and $\lambda_2 \in \hat{\mathcal{A}}_2$ by $\lambda_1(x) = \lambda(x \otimes 1)$ and $\lambda_2(y) = \lambda(1 \otimes y)$. Then $\lambda \mapsto (\lambda_1, \lambda_2)$ is a bijection $(\mathcal{A}_1 \otimes \mathcal{A}_2)^\wedge \to \hat{\mathcal{A}}_1 \times \hat{\mathcal{A}}_2$.

1.4.19. <u>Caution.</u> It is in general <u>not</u> the case that $(\mathcal{A}_1 \otimes \mathcal{A})^\wedge \to \hat{\mathcal{A}}_1 \times \hat{\mathcal{A}}_2$ is a homeomorphism when $(\mathcal{A}_1 \times \mathcal{A}_2)^\wedge$ is given its Z-topology and $\hat{\mathcal{A}}_1 \times \hat{\mathcal{A}}_2$ is given the product of the Z-topologies of $\hat{\mathcal{A}}_1$ and $\hat{\mathcal{A}}_2$. We will see this by example below.

1.4.20. <u>Remark.</u> It is clear from 1.4.17 that $f \otimes g \mapsto fg$ defines an isomorphism of $K[\{X_\alpha\}_{\alpha \in A}] \otimes K[\{Y_\beta\}_{\beta \in B}]$ with $K[\{X_\alpha\}_{\alpha \in A} \cup \{Y_\beta\}_{\beta \in B}]$. If we identify $K[\{X_\alpha\}_{\alpha \in A}]^\wedge$ with $\prod_{\alpha \in A} K$, etc., using 1.2.13 then clearly the bijection of 1.4.18 becomes just the obvious bijection

$\prod_{\gamma \in A \cup B} K \approx (\prod_{\alpha \in A} K) \times (\prod_{\beta \in B} K)$. In particular, referring back to 1.4.7 we see that to get the example mentioned in 1.4.19 we need only show that the Zariski topology for $K \times K$, regarded as $K[X, Y]$, is not necessarily the product of the weakest T_1-topology on K with itself. If K is finite, then K and $K \times K$ are both discrete, so in this case $K \times K$ does have the product topology. But now suppose K is infinite. A closed subset of $K \times K$ in the product topology is just a finite union of sets of the form $F_1 \times F_2$ where each of F_1 and F_2 is either a finite subset of K or all of K. If either or both of F_1 and F_2 are finite sets then clearly $F_1 \times F_2$ meets the diagonal Δ of $K \times K$ in a finite set, and no finite union of such sets can include Δ. Thus the only closed set of the product topology which includes Δ is $K \times K$; i.e., Δ is dense in $K \times K$ with respect to the product topology. But on the other hand Δ is clearly closed in the Z-topology since it is the zero set of the polynomial $X-Y$.

1.4.21. Definition. Given an algebra $\mathcal{A}$ over K and an extension field F of K we denote by $\hat{\mathcal{A}}_F$ the set $\mathrm{Hom}_K(\mathcal{A}, F)$ of algebra homomorphisms of $\mathcal{A}$ into F. For each $x \in \mathcal{A}$ we define $\hat{x} : \hat{\mathcal{A}}_F \to F$ by $\hat{x}(\varphi) = \varphi(x)$, and note that $x \mapsto \hat{x}$ is an algebra homomorphism $\mathcal{A} \to F^{\hat{\mathcal{A}}_F}$. We define the Z-topology for $\hat{\mathcal{A}}_F$ to be the weakest topology making all the $\hat{x}$ continuous (when F has its weakest T_1-topology) or equivalently the topology whose closed sets are generated by sets of the form $\hat{x}^{-1}(0)$.

1.4.22. Remark. Note that $\hat{\mathcal{A}} = \mathrm{Hom}_K(\mathcal{A}, K) = \hat{\mathcal{A}}_K$. Since K is a subfield of F we have $\hat{\mathcal{A}} = \hat{\mathcal{A}}_K \subseteq \hat{\mathcal{A}}_F$. Moreover, clearly for $x \in \mathcal{A}$ we have $\hat{x} : \hat{\mathcal{A}}_F \to F$ extending the previously defined $\hat{x} : \hat{\mathcal{A}}_K \to K$, and it follows that $\hat{\mathcal{A}}$ with its Z-topology is a subspace of $\hat{\mathcal{A}}_F$ with its Z-topology. The points of $\hat{\mathcal{A}}$ are frequently called "rational" points to distinguish them from more general elements of $\hat{\mathcal{A}}_F$.

1.4.23. Remark. $\hat{F}_F = \mathrm{Hom}_K(F, F)$ is just the Galois group of F over K, i.e., the group of automorphisms of F leaving points of K fixed. Note that this group, G, acts naturally on $\hat{\mathcal{A}}_F$. Given $g \in G$ and $\varphi \in \hat{\mathcal{A}}_F$, $g \circ \varphi \in \hat{\mathcal{A}}_F$. Clearly $\hat{\mathcal{A}}$ is left fixed by all elements of G, and in fact if F is a Galois extension then $\hat{\mathcal{A}}$ is just the fixed set of G. Note in particular the case $K = \mathbb{R}$, $F = \mathbb{C}$. In this case $G \simeq \mathbb{Z}_2$ (the identity and complex conjugation). Given an algebra $\mathcal{A}$ over $\mathbb{R}$ and $\varphi \in \hat{\mathcal{A}}_{\mathbb{C}}$ we define $\overline{\varphi} \in \hat{\mathcal{A}}_{\mathbb{C}}$ by $\overline{\varphi}(x) = \overline{\varphi(x)}$, so $\hat{x}(\overline{\varphi}) = \overline{\hat{x}(\varphi)}$ and $\hat{\mathcal{A}} = \hat{\mathcal{A}}_{\mathbb{R}}$ is the set of $\varphi \in \hat{\mathcal{A}}_{\mathbb{C}}$ such that $\varphi = \overline{\varphi}$.

1.4.24. Remark. We have an identification $K[\{X_\alpha\}_{\alpha \in A}]^\wedge_F \leftrightarrow \prod_{\alpha \in A} F$ just as in 1.2.13, so that the inclusion of $K[\{X_\alpha\}]^\wedge$ is just the obvious inclusion of $\prod_{\alpha \in A} K$ in $\prod_{\alpha \in A} F$. If $\{x_\alpha\}_{\alpha \in A}$ is a set of generators for $\mathcal{A}$ and $h : K[\{X_\alpha\}_{\alpha \in A}] \to \mathcal{A}$ is the surjective homomorphism defined by $h(X_\alpha) = x_\alpha$, and $\mathcal{I} = \ker(h)$, then we have noted in 1.4.15 the image of $\hat{h} : \hat{\mathcal{A}} \to K[\{X_\alpha\}_{\alpha \in A}]^\wedge = \prod_{\alpha \in A} K$ is just the set of $\{\xi_\alpha\}$ such that $P(\xi_\alpha) = 0$ for all $P \in \mathcal{I}$. Now $\hat{h}$ clearly extends to an injection of $\hat{\mathcal{A}}_F$ into

$K[\{X_\alpha\}_{\alpha\in A}]^\wedge_F = \prod_{\alpha\in A} F$ and again the image is the set of $\{\xi_\alpha\}$ such that $P(\xi_\alpha) = 0$ for all P in $\mathcal{I}$.

1.4.25. Remark. An alternative and useful way to look at $\hat{\mathcal{A}}_F$ is to "extend the ground field" from K to F. Given the K-algebra $\mathcal{A}$, $\mathcal{A} \otimes_K F$, includes both $\mathcal{A}$ and F as subalgebras (via the injections $x \mapsto x \otimes 1$ and $\alpha \to 1 \otimes \alpha$; since $k \otimes 1 = 1 \otimes k$ for $k \in K$ the copies of K in $\mathcal{A}$ and F are the same). In particular since $\mathcal{A} \otimes_K F$ includes F it can be regarded as an F-algebra. Now recall that we have a bijection of $\mathrm{Hom}_K(G, F) \times \mathrm{Hom}_K(F, F)$ with $\mathrm{Hom}_K(\mathcal{A} \otimes_K F, F)$, given by $(\varphi, g) \to \varphi \otimes g$. In particular taking $g = \mathrm{id} : F \to F$ gives a biejction of $\hat{\mathcal{A}}_F = \mathrm{Hom}_K(\mathcal{A}, F)$ with the set of K-algebra homomorphism $\psi : \mathcal{A} \otimes_K F \to F$ which are the identity on F. But the latter is just the set of F-algebra homomorphisms of $\mathcal{A} \otimes_K F$ into F. In other words, $\hat{\mathcal{A}}_F$ is just the set of rational points of $\mathcal{A} \otimes_K F$ considered as an F-algebra.

1.5. Ringed Spaces.

1.5.1. Definition. A structure ring (over K) for a set S is a subalgebra $\mathcal{A}$ of the algebra K^S of K valued functions on S which "separates points" (i.e., given $s_1, s_2 \in S$ with $s_1 \neq s_2$ there is an $f \in \mathcal{A}$ such that $f(s_1) \neq f(s_2)$). A ringed space over K consists of a pair $(S, \mathcal{A})$ where S is a set and $\mathcal{A}$ is a structure ring for S. If $(S_1, \mathcal{A}_1)$ and $(S_2, \mathcal{A}_2)$ are ringed spaces over K a morphism $(S_1, \mathcal{A}_1) \to (S_2, \mathcal{A}_2)$ (of ringed spaces) is a set mapping $\varphi: S_1 \to S_2$ such that $K^{\varphi}: K^{S_2} \to K^{S_1}$ maps $\mathcal{A}_2$ into $\mathcal{A}_1$, i.e., such that $f \circ \varphi \in \mathcal{A}_1$ whenever $f \in \mathcal{A}_2$. In this case we shall usually write φ^* for the homomorphism $f \mapsto f \circ \varphi$ of $\mathcal{A}_2$ into $\mathcal{A}_1$.

1.5.2. Notations and Conventions. Just as one frequently refers to a topological space by naming its underlying point set, so we shall normally speak of "the ringed space S", rather than "the ringed space $(S, \mathcal{A})$. In such contexts we shall use $\mathcal{A}(S)$ to refer to the structure ring of S. Also as above we will frequently drop explicit reference to the ground field K.

For each $s \in S$ we denote by $Ev(s) : \mathcal{A} \to K$ the map $f \mapsto f(s)$. The condition on a structure ring that it separates points is precisely equivalent to the statement that $Ev : S \to \hat{\mathcal{A}}$ is injective. It is frequently a considerable conceptual simplification to regard Ev as an identification of S with a subset of $\hat{\mathcal{A}}$, and where convenient we shall do so, often tacitly. Note that with this understanding we have $f = \hat{f}|S$, where $f \mapsto \hat{f}$ is the Gelfand representation of $\mathcal{A}$.

1.5.3. Definition. If $(S, \mathcal{A})$ is a ringed space and S' is a subset of S we make S' into a ringed space, called a ringed subspace of S, by defining its structure ring to be the algebra of functions on S' which are restrictions to S' of functions in the structure ring $\mathcal{A}$ of S.

1.5.4. Definition. If $(S, \mathcal{A})$ is a ringed space then subsets of S of the form $S_f = \{s \in S | f(s) \neq 0\}$ are called basic open sets. The Z-topology for S is the topology for S induced from the Z-topology for $\hat{\mathcal{A}}$, or more precisely the topology for S making $Ev : S \to \hat{\mathcal{A}}$ a homeomorphism into.

1.5.5. Proposition. If $(S, \mathcal{A})$ is a ringed space then the Z-topology of S can be characterized in any of the following equivalent ways:

1) The topology for S whose closed sets are of the form $\{s \in S | f(s) = 0$ for all $f \in \mathcal{I}\}$ where $\mathcal{I}$ is a subset of $\mathcal{A}$ (which of course can be taken to be an ideal and even a strict radical ideal).

2) The weakest topology for S such that each of the maps $f : S \to K$, $f \in \mathcal{A}$, is continuous when K is given its weakest T_1-topology.

3) The topology for S such that $B \subseteq S$ is closed if and only if for each $s \notin B$ there exists $f \in \mathcal{A}$ which vanishes on B but does not vanish at s, or equivalently the topology for S such that the closure of any $B \subseteq S$ is the set of $s \in S$ such that $f \in \mathcal{A}$ and $f|B = 0$ implies $f(s) = 0$.

4) The topology for S whose open sets are those sets $\mathcal{O}$ such that given $x \in \mathcal{O}$ there exists $f \in \mathcal{A}$ with $x \in S_f \subseteq \mathcal{O}$ (where S_f is the basic open set $\{s \in S \mid f(s) \neq 0\}$.

<u>Proof.</u> Cf., 1.4.8, 1.4.12, and 1.4.13. ∎

1.5.6. <u>Proposition.</u> If $f : (S_1, \mathcal{A}_1) \to (S_2, \mathcal{A}_2)$ is a morphism of ringed spaces the f is continuous with respect to the Z-topologies. Moreover, the homomorphism $f^* : \mathcal{A}_2 \to \mathcal{A}_1$ $(g \mapsto g \circ f)$ is injective if and only if $f(S_1)$ is Z-dense in S_2.

Proof. Since $f : S_1 \to S_2$ is clearly the restriction of $\hat{f}^* : \hat{\mathcal{A}}_1 \to \hat{\mathcal{A}}_2$, the continuity of f follows from 1.4.14. Now f^* is injective if and only if for all $g \in \mathcal{A}_2$, $g \circ f = 0 \Rightarrow g = 0$, i.e., $g(f(s)) = 0$ for all $s \in S_1$ implies $g = 0$, i.e., $g|f(S_1) = 0$ implies $g = 0$. By 3) of 1.5.5 this is equivalent to $f(S_1)$ being Z-dense in S_2. ∎

1.5.7. Proposition. If S_0 is a ringed subspace of S_1 then S_0 with its Z-topology is a topological subspace of S_1 with its Z-topology.

Proof. Immediate from 2) of 1.5.5. ∎

1.5.8. Proposition. If S_0 is a ringed subspace of the ringed space S_1 then the following are equivalent:

(1) S_0 is Z-dense in S_1 in the Z-topology.

(2) $f \mapsto f|S_0$ maps $\mathcal{A}(S_1)$ isomorphically onto $\mathcal{A}(S_0)$, i.e., every element of $\mathcal{A}(S_0)$ extends uniquely to an element of $\mathcal{A}(S_1)$.

Proof. By definition of ringed subspace, $f \mapsto f|S_0$ maps $\mathcal{A}(S_1)$ onto $\mathcal{A}(S_0)$. If $i : S_0 \to S_1$ is the inclusion map then $f|S_0 = f \circ i$. The theorem now follows from 1.5.6. ∎

1.5.9. Proposition. If $(S, \mathcal{A})$ is a ringed space and $(S_0, \mathcal{A}_0)$ is a ringed subspace, then $\hat{\mathcal{A}}_0$ is the Z-closure of S_0 in $\hat{\mathcal{A}}$. More precisely, if ρ denotes the restriction homomorphism $f \mapsto f|S_0$ of $\mathcal{A}$ onto $\mathcal{A}_0$ then the injection $\hat{\rho} : \hat{\mathcal{A}}_0 \to \hat{\mathcal{A}}$ is a Z-homeomorphism of $\hat{\mathcal{A}}_0$ onto the Z-closure of S_0 in $\hat{\mathcal{A}}_0$.

Proof. Let $\mathcal{I} = \{f \in \mathcal{A} \mid (f|S_0) = 0\}$, so $\mathcal{I}$ is the kernel of ρ. Then $V(\mathcal{I}) = \{\varphi \in \hat{\mathcal{A}} \mid \hat{f}(\varphi) = 0$ for all $f \in \mathcal{A}\}$ is the Z-closure of S_0 in $\hat{\mathcal{A}}$. Now we have a unique isomorphism $\mathcal{A}/\mathcal{I} \simeq \mathcal{A}_0$ commuting with $\rho : \mathcal{A} \to \mathcal{A}_0$ and $\pi : \mathcal{A} \to \mathcal{A}/\mathcal{I}$ and the proposition follows directly from 1.4.14. ∎

1.5.10. Definition. A ringed space $(S, \mathcal{A})$ is called complete if $S = \hat{\mathcal{A}}$ (or more precisely if $Ev : S \to \hat{\mathcal{A}}$ is surjective). A completion of a ringed space S_0 is a complete ringed space S such that S_0 is a Z-dense ringed subspace of S.

1.5.11. Proposition. The category of complete ringed spaces over K is (isomorphic to) the dual of the category of strictly semi-simple algebras over K.

Proof. Exercise.

1.5.12. Proposition. If S is a complete ringed space then so is any ringed subspace S_0 which is Z-closed in S. Moreover, if S is a ringed subspace of a ringed space S_1 then S is Z-closed in S_1.

Proof. Immediate from 1.5.9. Note that if S is a ringed subspace of $(S_1, \mathcal{A}_1)$ then by 1.5.9, S is actually Z-closed in $\hat{\mathcal{A}}_1$ and a fortiori Z-closed in S_1. ∎

1.5.13. Proposition. If $(S, \mathcal{A})$ is a ringed space then $(\hat{\mathcal{A}}, \mathcal{A})$ is a completion of S. If $(S', \mathcal{A}')$ is any completion of S then the identity map of S extends uniquely to a ringed space isomorphism of S' with $\hat{\mathcal{A}}$.

Proof. We are of course identifying S with its image in $\hat{\mathcal{A}}$ under $Ev : S \to \hat{\mathcal{A}}$ and regarding $\mathcal{A}$ as an algebra of functions on $\hat{\mathcal{A}}$ via the Gelfand representation $f \mapsto \hat{f}$. Since $f = \hat{f}|S$, S is dense in $\hat{\mathcal{A}}$ by 1.5.8. Since $g \mapsto g|S$ maps $\mathcal{A}'$ isomorphically onto $\mathcal{A}$ we can identify $\mathcal{A}'$ with $\mathcal{A}$ so that S is a ringed subspace of S' which in turn is a ringed subspace of $\hat{\mathcal{A}}$. Since S is dense in $\hat{\mathcal{A}}$, and S' is closed in $\hat{\mathcal{A}}$ by 1.5.12, $S' = \hat{\mathcal{A}}$. ∎

1.5.14. Definition. If K is a topological field and $(S, \mathcal{A})$ is a ringed space over K then the W-topology for S is the weakest topology for S such that each of the maps $f : S \to K$, $f \in \mathcal{A}$ is continuous.

1.5.15. Proposition. If K is a topological field and $(S, \mathcal{A})$ is a ringed space over K then the W-topology for S is always Hausdorff and at least as strong as the Z-topology for S (which of course need not be Hausdorff). A necessary and sufficient condition for the Z-topology and the W-topology to be the same is that given a W-closed subset F of S and $s_0 \in S-F$ there exist an $f \in \mathcal{A}$ such that $f|F = 0$ and $f(s_0) \neq 0$.

Proof. Since K is Hausdorff and $\mathcal{A}$ separates points of S it is clear that the W-topology for S is Hausdorff. That the W-topology is at least as strong as the Z-topology and the condition that they agree follow directly from 1.4.9 and 1.5.5. ∎

1.5.16. Definition. If S is a ringed space over K then a function $h : S \to K$ will be called regular if it can be written in the form f/g where $f, g \in \mathcal{A}(S)$ and g never vanishes on S. The ring of regular functions on S will be denoted by $\mathcal{A}_{reg}(S)$ and the ringed space $(S, \mathcal{A}_{reg}(S))$ will be called the regularization of S. We call S a regular ringed space if $\mathcal{A}(S) = \mathcal{A}_{reg}(S)$; i.e., if given $f \in \mathcal{A}(S)$ such that f never vanishes on S it follows that $1/f \in \mathcal{A}(S)$.

1.5.17. Remark. Suppose S is a ringed space over K and $h \in \mathcal{A}_{reg}(S)$, say $h = f/g$ with $f, g \in \mathcal{A}(S)$ and g never zero. Then clearly $h^{-1}(0) = f^{-1}(0)$. It follows that the Z-topology for S remains the same when we regularize. Similarly if K is a topological field, then $x \mapsto 1/x$ is continuous from $K-\{0\}$ to K, so that since $g : S \to K-\{0\}$ is continuous, so is $1/g$ and hence f/g. It follows that the W-topology for S also does not change when we regularize.

1.5.18. Lemma. Let $(S, \mathcal{A})$ be a ringed space and let $\varphi \in \hat{\mathcal{A}}$. If $\varphi \notin S$ then for each $s \in S$ there exists $f \in \mathcal{A}$ such that $f(s) \neq 0$ and $\varphi(f) = \hat{f}(\varphi) = 0$.

Proof. If not there would exist $s \in S$ such that $\ker(\varphi) \subseteq \ker(\varphi_s)$. Since $\ker(\varphi)$ and $\ker(\varphi_s)$ are strictly maximal ideals it follows that $\ker(\varphi) = \ker(\varphi_s)$ so by 1.2.3 $\varphi = \varphi_s$, i.e., $\varphi \in S$, a contradiction. ∎

1.5.19. Theorem. Let $(S, \mathcal{A})$ be a regular ringed space over $\mathbb{R}$ and suppose S is compact in the W-topology (or, only apparently more generally, in some topology stronger than the W-topology, so that in this topology all $f \in \mathcal{Q}$ are continuous). Then $(S, \mathcal{Q})$ is a complete ringed space.

Proof. Recall that in the lattice of topologies for a set, compact Hausdorff topologies are minimal with respect to Hausdorff topologies (i.e., a compact Hausdorff topology cannot be weakened and remain Hausdorff). It follows that we can assume it is the W-topology which is compact (cf., 1.5.15). If the theorem were false then we could find $\varphi \in \hat{\mathcal{Q}}$ such that $\varphi \notin S$, and by the lemma for each $s \in S$ we could find $f^s \in \mathcal{Q}$ such that $f^s(s) \neq 0$ and $\hat{f^s}(\varphi) = \varphi(f^s) = 0$. We can suppose f^s is everywhere non-negative (otherwise replace f^s by its square) and we let U_s be the set $\{x \in S \mid f^s(x) > 0\}$. Since f^s is continuous this is an open set containing s, and since S is compact we can find $s_1, \ldots, s_n$ such that $U_{s_1}, \ldots, U_{s_n}$ cover S. Then clearly $f = f^{s_1} + \ldots + f^{s_n}$ is positive everywhere on S so, since $(S, \mathcal{Q})$ is regular, it follows that $1/f \in \mathcal{Q}$. On the other hand, $\varphi(f) = \varphi(f^{s_1}) + \ldots + \varphi(f^{s_n}) = 0$ so $\varphi(1) = \varphi(f \cdot (1/f)) = \varphi(f)\varphi(1/f) = 0$, contradicting that $\varphi : \mathcal{Q} \to \mathbb{R}$ is a (unitary) homomorphism. ■

1.5.20. Corollary. If $(S, \mathcal{Q})$ is a ringed space over $\mathbb{R}$ and S is W-compact, then $(S, \mathcal{Q}_{reg}(S))$, the regularization of S, is a complete ringed space.

Proof. Cf. 1.5.17. ■

1.5.21. Remark. Let $\mathcal{Q}_1$ be a subalgebra of K^{S_1} and $\mathcal{Q}_2$ a subalgebra of K^{S_2}. Suppose $f_1, \ldots, f_n$ are linearly independent elements of $\mathcal{Q}_1$ and $g_1, \ldots, g_m$ are linearly independent elements of $\mathcal{Q}_2$ and suppose $\sum_{i,j} a_{ij} f_i(x) g_j(y) = 0$ for all $x, y \in S_1 \times S_2$, where $a_{ij} \in K$. Then for each $x \in S_1$ we have $\sum_j (\sum_i a_{ij} f_i(x)) g_j = 0$ so $\sum_j a_{ij} f_i(x) = 0$ for $j = 1, \ldots, m$ and so $\sum_i a_{ij} f_j = 0$, and all the $a_{ij} = 0$. In other words the functions $(x, y) \mapsto f_i(x) g_j(y)$ are linearly independent in $K^{S_1 \times S_2}$ It follows that we may identify $\mathcal{Q}_1 \otimes \mathcal{Q}_2$ with the subalgebra of $K^{S_1 \times S_2}$ consisting of functions which are finite sums of functions of the form $f \otimes g$ with $f \in \mathcal{Q}_1$ and $g \in \mathcal{Q}_2$, where $f \otimes g(x, y) = f(x) g(y)$.

1.5.22. Definition. Given ringed spaces S_1 and S_2 over K we make $S_1 \times S_2$ into a ringed space by defining

$$\mathcal{Q}(S_1 \times S_2) = \mathcal{Q}(S_1) \otimes \mathcal{Q}(S_2)$$

where (see 1.5.21) we identify $f \otimes g$ in $\mathcal{Q}(S_1) \otimes \mathcal{Q}(S_2)$ with the function $(x, y) \mapsto f(x) g(y)$. In particular $f \in \mathcal{Q}(S_1)$ is identified with $f \otimes 1 \in \mathcal{Q}(S_1 \times S_2)$, i.e., with $(x, y) \mapsto f(x)$ and similarly $g \in \mathcal{Q}(S_2)$ is identified with $(x, y) \mapsto g(y)$. It follows of course that $\mathcal{Q}(S_1)$ and $\mathcal{Q}(S_2)$ are canonically included as subalgebras of $\mathcal{Q}(S_1 \times S_2)$ and together they generate $\mathcal{Q}(S_1 \times S_2)$. Hence for example if $\mathcal{Q}(S_1)$ and $\mathcal{Q}(S_2)$ are both finitely generated as K-algebras so is $\mathcal{Q}(S_1 \times S_2)$.

1.5.23. Remark. If Π_1 and Π_2 are the projection maps of

$S_1 \times S_2$ on S_1 and S_2 respectively then clearly $\Pi_1^* : \mathcal{Q}(S_1) \to \mathcal{Q}(S_1 \times S_2)$ is just $f \mapsto f \otimes 1$, the above identification and similarly Π_2^* is $g \mapsto 1 \otimes g$. In particular Π_1 and Π_2 are ringed space morphisms. Given $s_1 \in S_1$ the natural "inclusion" map $i_{s_1} : S_2 \to S_1 \times S_2$, $s_2 \mapsto (s_1, s_2)$ is also clearly a ringed space morphism. In fact $i_{s_1}^*(f \otimes g) = f(s_1)g$. If we regard its image $\{s_1\} \times S_2$ as a subspace of $S_1 \times S_2$ it is Z-closed (it equals $\Pi_2^{-1}(s_1)$) and is isomorphic as a ringed space to S_2, since Π_2 restricted to $\{s_1\} \times S_2$ is clearly inverse to i_{s_1}. Given $\varphi : S \to S_1 \times S_2$, where S is some ringed space over K, say $\varphi(s) = (\varphi_1(s), \varphi_2(s))$, then if φ is a ringed space morphism it follows that $\varphi_i = \Pi_i \circ \varphi$ is a ringed space morphism $(i = 1, 2)$; conversely if φ_1 and φ_2 are ringed space morphisms then given $f \in \mathcal{Q}(S_1)$ and $g \in \mathcal{Q}(S_2)$ we have

$$\varphi^*(f \otimes g)(s) = f \otimes g(\varphi(s)) = f \otimes g(\varphi_1(s), \varphi_2(s)) = f(\varphi_1(s))g(\varphi_2(s))$$

or $\varphi^*(f \otimes g) = \varphi_1^*(f)\varphi_2^*(g) \in \mathcal{Q}(S)$. Thus φ is a ringed space morphism if and only if φ_1 and φ_2 are, so that $S_1 \times S_2$ is the product of S_1 and S_2 in the category of ringed spaces.

1.5.24. Proposition. The product $S_1 \times S_2$ of ringed spaces over K is complete if and only if S_1 and S_2 are each complete. In general the completion of $S_1 \times S_2$ is the product of the completions of S_1 and S_2.

Proof. Immediate from 1.4.18.

1.5.25. Remark. In general the Z-topology for $S_1 \times S_2$ is not the product of the Z-topology for S_1 and the Z-topology for S_2, but rather a much finer topology (cf., 1.4.20). However, suppose K is a topological field. Since functions of the form $f \otimes g : (s_1, s_2) \mapsto f(s_1)g(s_2)$ generate $\mathcal{A}(S_1 \times S_2)$, the W-topology for $S_1 \times S_2$ (cf., 1.5.14) is the weakest topology making all such $f \otimes g : S_1 \times S_2 \to K$ continuous. But since multiplication is a continuous map $K \times K \to K$ and $f \otimes g = (f \otimes 1)(1 \otimes g)$, the W-topology for $S_1 \times S_2$ is the weakest topology such that each $f \otimes 1$ and $1 \otimes g$ is continuous $S_1 \times S_2 \to K$. But this shows that the W-topology for $S_1 \times S_2$ is the product of the W-topology for S_1 and the W-topology for S_2. Sums also exist in the category of ringed spaces over K. Given $(S_1, \mathcal{A}_1)$ and $(S_2, \mathcal{A}_2)$, ringed spaces over K, their sum in the category of ringed spaces over K is $(S_1+S_2, \mathcal{A}_1 \oplus \mathcal{A}_2)$ where S_1+S_2 denotes the disjoint union of S_1 and S_2 and for $f_1 \in \mathcal{A}_1$, $f_2 \in \mathcal{A}_2$, $f_1 \oplus f_2$ equals f_1 on S_1 and equals f_2 on $\mathcal{A}_2$. S_1+S_2 is complete if and only if each of S_1 and S_2 is complete and in general the completion of S_1+S_2 is the sum of the completions of S_1 and of S_2. The Z-topology for S_1+S_2 is the sum of the Z-topologies for S_1 and S_2 and if K is a topological field the same is true for the W-topologies. These facts follow easily from 1.4.16.

1.5.26. Notation. If G is a group and S is a set then an action of G on S is a map $\alpha : G \times S \to S$ such that for each $g \in G$, $s \mapsto \alpha(g, s)$ is a bijection α_g of S with itself and $g \mapsto \alpha_g$ is a homomorphism of G into the group of bijections of S. A G-set is a pair consisting of a set S and a fixed action α of G on S. In general we write simply gs instead of $\alpha(g, s)$.

1.5.27. Definition. A ringed group over K is a ringed space $(G, \mathcal{A}(G))$ over K whose underlying set is a group in such a way that the map $(x, y) \mapsto xy^{-1}$ of $G \times G \to G$ is a morphism of ringed spaces. Ringed subgroup and homomorphism of ringed groups have the obvious meanings. A ringed space $(X, \mathcal{A}(X))$ whose underlying set is a G-set is called a ringed G-space if the map $(g, x) \mapsto gx$ of $G \times X \to X$ (i.e., the action of G on X) is a ringed space morphism.

1.5.28. Definition. Let K be an infinite field. If X is a ringed space over K then a one parameter group of automorphisms of X is an action $\varphi : K \times X \to X$ (say, $(t, x) \mapsto \varphi_t(x)$) such that there exists a homomorphism $\Phi : \mathcal{A}(X) \to \mathcal{A}(X)[T]$ satisfying $f \circ \varphi_t = \Phi(f)(t)$ for all $f \in \mathcal{A}(X)$ and all $t \in K$. (We note that Φ is unique, for if $\Psi : \mathcal{A}(X) \to \mathcal{A}(X)[T]$ satisfied the same property then for $f \in \mathcal{A}(X)$ we would have $\Psi(f)-\Phi(f) = \sum_{i=0}^{n} g_i T^i$, $g_i \in \mathcal{A}(X)$. Then for any $t \in K$ and $x \in X$ we would have $\sum_{i=0}^{n} g_i(x)t^i = f \circ \varphi_t(x) - f \circ \varphi_t(x) = 0$, and since then are infinitely many t in K, $g_i(x) = 0$ $(i = 1, \ldots, n)$, so $\Psi(f) = \Phi(f)$).

1.5.29. Remark. We now try to motivate 1.5.28. The field K has a natural structure of ringed space, the structure ring being the polynomial ring $K[T]$ (where $P(T) \in K[T]$ is identified with the function $\alpha \mapsto P(\alpha)$; cf., 1.2.11). If X is any ringed space over K then

$$\mathcal{A}(K \times X) = \mathcal{A}(K) \otimes \mathcal{A}(X) = K[T] \otimes \mathcal{A}(X) = \mathcal{A}(X)[T],$$

that is, elements f of the structure ring of $K \times X$ are of the form $(t,x) \mapsto \sum_{i=0}^{n} g_i(x)t^i$ for $g_i \in \mathcal{A}(X)$. In particular $\mathcal{A}(K \times K) = K[X, Y]$ the polynomial functions on $K \times K$. If $P(T) \in K[T]$ then $P(X-Y) \in K[X, Y]$; hence K is a ringed group under addition. Let $\varphi : K \times X \to X$ be an action of K on X, $(t,x) \mapsto \varphi_t(x)$. We leave it to the reader to check that φ is a one parameter group of automorphisms of X precisely when φ is a ringed space morphism, i.e., when X is a ringed K-space.

1.5.30. Remark. Let $(S, \mathcal{A})$ be a ringed space. Given a Z-closed subset X of S let $\widetilde{X}$ denote the Z-closure of X in $\hat{\mathcal{A}}$. Then $X \mapsto \widetilde{X}$ is clearly a bijection correspondence between the collection of Z-closed subsets of S and those Z-closed subsets F of $\hat{\mathcal{A}}$ such that $F \cap S$ is Z-dense in F. Now clearly $f \in \mathcal{A}$ vanishes on X if and only if f vanishes on $\widetilde{X}$, so that $I(X) = I(\widetilde{X})$. It follows from 1.3.9 that $X \mapsto I(X)$ is an inclusion reversing bijective correspondence between the collection of Z-closed subsets of S and a certain set of strict radical ideals of $\mathcal{A}$ (namely, those strict radical ideals I whose variety $V = V(I)$ satisfies $V \cap S$ is Z-dense in V).

1.5.31. Definition. Let X be a ringed space, $f \in \mathcal{A}(X)$, and let $X_f = \{x \in X \mid f(x) \neq 0\}$ be the basic open set defined by f (cf., 1.5.3) considered as a ringed subspace of X. Let $\widetilde{X}_f$ denote the regularization of X_f (1.5.16). We define a ringed space $X_{(f)}$ with the same underlying point set as X_f (and $\widetilde{X}_f$) by defining its structure ring to be

$$\mathcal{A}(X_{(f)}) = \mathcal{A}(X_f)[1/f] \subseteq \mathcal{A}(\widetilde{X}_f),$$

i.e., a function $g : X_f \to K$ is in the structure ring of $X_{(f)}$ if and only if it can be expressed in the form

$$x \mapsto h_0(x)+h_1(x)/f(x) + \ldots + h_n(x)/f(x)^n$$

with $h_1, \ldots, h_n \in \mathcal{Q}(X)$. We note that since $\mathcal{Q}(X_f) \subseteq \mathcal{Q}(X_{(f)}) \subseteq \mathcal{Q}(\tilde{X}_f)$ the identity maps $\tilde{X}_f \to X_{(f)} \to X_f$ are ringed space morphisms.

1.5.32. Remark. If X is a ringed space, and $\mathcal{O}$ a basic open set of X there will always by many f in $\mathcal{Q}(X)$ such that $\mathcal{O} = X_f$. For example, f can be replaced by any of its powers. Different choices of f will generally give different ringed space structures $X_{(f)}$ to $\mathcal{O}$ and in general there is no canonical choice for f. However, in one important class of cases there is a canonical "strongest" choice for f. Namely, let us call $\mathcal{O}$ a principal open set of X if the Z-closed set $V = X - \mathcal{O}$ has for its ideal $\mathcal{J}(V)$ a principal ideal of $\mathcal{Q}(X)$, say $\mathcal{J}(V) = (P)$. If also $\mathcal{J}(V) = (Q)$ then $P = QS_1$ and $Q = PS_2$ where $S_1, S_2 \in \mathcal{Q}(X)$. Now in $\mathcal{O}$, where neither P nor Q vanishes we have $(1/P) = S_2(1/Q)$ and $(1/Q) = S_1(1/P)$ so that $\mathcal{Q}(\mathcal{O})[1/P] = \mathcal{Q}(\mathcal{O})[1/Q]$ and hence $X_{(P)} = X_{(Q)}$.

1.5.33. Proposition. If X is a complete ringed space and $f \in \mathcal{Q}(X)$ then $X_{(f)}$ is a complete ringed space.

Proof. Recall (1.2.11) that K is a complete ringed space over K whose structure ring $\mathcal{P}(K)$ is generated by the identity map $K \to K$. Thus $X \times K$ is a complete ringed space (1.5.24) and hence so is its Z-closed subspace $V = \{(x,t) \in X \times K \mid tf(x) = 1\}$ (1.5.12). Now the projection $\prod : X \times K \to X$ is a ringed space morphism (1.5.23); hence so is its restriction to V. Now clearly $\Pi(V) = X_f$ and $\psi : X_{(f)} \to V$, $x \mapsto (x, 1/f(x))$ is a ringed space morphism inverse to $\Pi | V$. Thus $X_{(f)}$ is isomorphic to V and hence complete. ■

1.6. Ringed Space Categories.

We shall call a category $\mathcal{C}$ a category of structured sets if we are given some weakening of structure ("forgetful") functor from $\mathcal{C}$ to the category of sets. Thus to each object X of $\mathcal{C}$ we have associated a set, called the underlying set of X (which we usually denote simply by X when there is no ambiguity). And given a second object Y of $\mathcal{C}$ the set $\mathcal{C}(X, Y)$ of morphisms of X to Y injects into the set of set mappings of X to Y, with composition of morphisms going over to the usual composition of set maps. Of course, most of the standard categories of mathematics are of this sort, e.g., groups, rings, topological spaces, ringed spaces.

1.6.1. Definition. A ringed space category over K is a category $\mathcal{C}$ of structured sets together with a function which associates to each object X of $\mathcal{C}$ a structure ring $\mathcal{C}(X)$ for the underlying set of X, such that given objects X and Y of $\mathcal{C}$, a set mapping $h : X \to Y$ between their underlying sets is a morphism of $\mathcal{C}$ if and only if it is a ringed space morphism $(X, \mathcal{C}(X)) \to (Y, \mathcal{C}(Y))$ (i.e., if and only if $g \circ h \in \mathcal{C}(X)$ whenever $g \in \mathcal{C}(Y)$).

Of course, for any K the category of all ringed spaces over K, or any of its full subcategories, is a ringed space category over K. What makes the notion of ringed space category interesting is that many familiar categories, while not ringed space categories by definition, can be given natural structures of ringed space category and so can be regarded as full subcategories of the category of ringed spaces (over some K). This allows one to treat seemingly diverse questions in a uniform way and suggests techniques for studying such categories that might not otherwise be evident. We give some examples below.

1.6.2. Example. Let $\mathcal{C}$ denote the category of completely regular topological spaces and continuous maps between them. For each object X of $\mathcal{C}$ let $\mathcal{C}(X)$ denote the algebra (over $\mathbb{R}$) of bounded, continuous real valued functions on X. It is trivial of course (from the definition of complete regularity) that $\mathcal{C}(X)$ separates points of X and hence that $(X, \mathcal{C}(X))$ is a ringed space over $\mathbb{R}$. Moreover,

(*) If X is a completely regular space, then the topology for X is the W-topology of $(X, \mathcal{C}(X))$, which is also the Z-topology for $(X, \mathcal{C}(X))$.

Indeed, that the topology for X is the W-topology for $(X, \mathcal{C}(X))$ is just the fact that the topology for X is the weakest with the given set of bounded, continuous real valued functions, a property well known to characterize completely regular topologies among all topologies. That the Z-topology is the same as the W-topology now follows directly from 1.5.15 and the Tietze extension theorem.

It is natural to ask under what conditions $(X, \mathcal{C}(X))$ is a complete ringed space, and more generally to characterize the completion of $(X, \mathcal{C}(X))$. In fact $\mathcal{C}(X)^\wedge$ with its W-topology is, essentially by definition, the Stone-Čech compactification of X. Thus $(X, \mathscr{C}(X))$ is complete if and only if X is compact and in general the completion of $(X, \mathscr{C}(X))$ is $(\overline{X}, \mathscr{C}(\overline{X}))$ where $\overline{X}$ is the Stone-Čech compactification of X.

If $\{y_\alpha\}$ is a net in the completely regular space Y then, by (*), $y_\alpha \to y$ if and only if $g(y_\alpha) \to g(y)$ for all $g \in \mathscr{C}(Y)$. If X is another completely regular space and $f : X \to Y$ is a set map such that $g \circ f$ is in $\mathscr{C}(X)$ for all $g \in \mathscr{C}(Y)$, then if $x_\alpha \to x$ in X it follows that $g \circ f(x_\alpha) \to g \circ f(x)$ for all $g \in \mathscr{C}(Y)$ and hence that $f(x_\alpha) \to f(x)$ in Y; so f is continuous (i.e., a morphism of $\mathscr{C}$). Of course, conversely if f is continuous, then $g \circ f \in \mathscr{C}(X)$ for all $g \in \mathscr{C}(Y)$ and it follows that the assignment $X \mapsto \mathscr{C}(X)$ does in fact make $\mathscr{C}$ into a ringed space category over $\mathbb{R}$. Note that $(X, \mathscr{C}(X))$ is not in general a regular ringed space. For example, taking $X = \mathbb{R}$, $x \mapsto (1/(1+x^2))$ is clearly in $\mathscr{C}(\mathbb{R})$ and never vanishes, but $x \mapsto (1+x^2)$ is unbounded so not in $\mathscr{C}(\mathbb{R})$. On the other hand, if X is compact and $f \in \mathscr{C}(X)$ never vanishes, then of course f is bounded away from zero, so $1/f \in \mathscr{C}(X)$; i.e., when X is compact $(X, \mathscr{C}(X))$ is a regular ringed space. Using 1.5.19 we have an alternative proof that $(X, \mathscr{C}(X))$ is a complete ringed space when X is compact.

1.6.3. Remark. Before giving further examples of ringed space categories, we will explain a simple but useful method for defining such structures.

Suppose $\mathscr{C}$ is a category of structured sets with a special object K whose underlying set is in fact the underlying set of the field K. Suppose also:

1) For every object X of $\mathcal{C}$, $\mathcal{C}(X, K)$ is a structure ring for (the underlying set of) X over K.

2) A set mapping $f : X \to Y$ between (the underlying sets of) objects of $\mathcal{C}$ is a morphism of $\mathcal{C}$ provided $g \circ f : X \to K$ is a $\mathcal{C}$-morphism whenever $g : Y \to K$ is.

Then clearly we can give $\mathcal{C}$ the structure of a ringed space category over K by defining $\mathcal{C}(X) = \mathcal{C}(X, K)$.

In practice condition 1) is often verified by checking the following five conditions which clearly imply it:

(1a) The product $K \times K$ exists in $\mathcal{C}$.

(1b) For each $\alpha \in K$ the map $\beta \mapsto \alpha\beta$ is a morphism $K \to K$ of $\mathcal{C}$.

(1c) The map $(\alpha, \beta) \mapsto \alpha+\beta$ is a morphism $K \times K \to K$ of $\mathcal{C}$.

(1d) The map $(\alpha, \beta) \mapsto \alpha\beta$ is a morphism $K \times K \to K$ of $\mathcal{C}$.

(1e) If X is an object of $\mathcal{C}$ and x_0, x_1 are distinct elements of the underlying set of X then there exists a $\mathcal{C}$-morphism $f : X \to K$ with $f(x_0) = 0$ and $f(x_1) = 1$.

The most obvious case of the above is the category $\mathcal{C}$ of <u>all</u> ringed spaces over K. We take as the "special object" the ringed space $(K, \mathcal{P}(K))$ where $\mathcal{P}(K)$ as usual denotes the ring of polynomial functions on K. We leave it to the reader to verify that if $(S, \mathcal{Q})$ is any ringed space over K then $\mathcal{Q}$ is in fact the set of ringed space morphisms $f : (S, \mathcal{Q}) \to (K, \mathcal{P}(K))$, i.e., that a function $f : S \to K$ is in $\mathcal{Q}$ if and only if $g \circ f \in \mathcal{Q}$ whenever $g \in \mathcal{P}(K)$ (cf., 2.0.5).

We now give some more interesting examples.

1.6.4. Example. In what follows k denotes either a positive integer or one of the symbols ∞ or ω. We denote by C^k the category of (paracompact) C^k manifolds, where as usual C^ω means real analytic. We shall take as well known the fact that every C^k manifold admits a C^k embedding in some $\mathbb{R}^n$. In the case $k = \omega$ this is a rather deep result, due to Grauert [11] and Morrey [22]. Of course, the field $\mathbb{R}$ of real numbers has a standard structure as object of C^k and for X a C^k manifold we put $C^k(X) = C^k(X, \mathbb{R})$. As an immediate consequence of the above embedding theorem it follows that if Y is a C^k manifold then:

(a) The elements of $C^k(Y)$ separate points of Y.

(b) Given $y_0 \in Y$ there exist $g_1, \ldots, g_n$ in $C^k(Y)$ forming a local coordinate system for Y at y_0.

Taking $K = \mathbb{R}$ and $\mathcal{C} = C^k$ it is easy to see that (1a)-(1e) of 1.6.3 above hold; the only non-trivial point is (1e) which is immediate from (a) above. Thus $C^k(Y)$ is a structure ring for Y. If X is a second C^k manifold and $f : X \to Y$ any function, then it is an easy consequence of (b) above that f is a C^k morphism if and only if $g \circ f \in C^k(X)$ whenever $g \in C^k(Y)$. Thus the assignment $Y \mapsto C^k(Y) = C^k(Y, \mathbb{R})$ does in fact give C^k the structure of a ringed space category over $\mathbb{R}$. The existence of a C^k embedding of a C^k manifold Y in some $\mathbb{R}^m$ implies in particular that the manifold topology of Y is the W-topology for $(Y, C^k(Y))$. If F is a closed subset of Y then for $k \neq \omega$ it is well known that for any $y_0 \in Y-F$ there exists a C^k function

$h : Y \to \mathbb{R}$ with $h|F = 0$ and $h(y_0) \neq 0$, so by 1.5.15 the manifold topology for Y is also the Z-topology for $(Y, C^k(Y))$ when $k \neq \omega$. For the case $k = \omega$ the situation is quite different. By the principle of analytic continuation, if an element f of $C^\omega(Y)$ vanishes in a neighborhood of a point y_0 then it vanishes on the connected component of y_0 in Y. Thus the Z-closure of a subset B of Y includes every connected component of Y which contains an interior point of B. Thus when Y has positive dimension it is clear that the Z-topology of $(Y, C^\omega(Y))$ is much weaker than the manifold topology. The Z-closed subsets of Y are usually called analytic subvarieties of Y.

The map $x \mapsto 1/x$ of $\mathbb{R}-\{0\} \to \mathbb{R}$ is a C^ω map and a fortiori a C^k map for all k. It follows that if $f \in C^k(Y)$ and f never vanishes then $1/f \in C^k(Y)$, i.e., $(Y, C^k(Y))$ is always a regular ringed space over $\mathbb{R}$. Then by 1.5.19 it follows that if Y is a compact C^k manifold then $(Y, C^k(Y))$ is complete.

1.6.5. Remark. As noted in 1.5.22 products always exist in the category of all ringed spaces. In a general ringed space category $\mathcal{C}$ products may or may not exist (e.g., products do not exist in the category of C^∞ manifolds with boundary), and when products do exist they are in general not the product in the category of all ringed spaces. The fact that the projections of $X \times Y$ on X and Y are in the category imply that $\mathcal{C}(X) \otimes \mathcal{C}(Y) \subseteq \mathcal{C}(X \times Y)$; however, in general $\mathcal{C}(X \times Y)$ will be some sort of completion of $\mathcal{C}(X) \otimes \mathcal{C}(Y)$. For example, if X and Y are compact spaces then the Stone-Weierstrass theorem implies that $C(X) \otimes C(Y)$ is dense in $C(X \times Y)$ in the compact open topology.

1.6.6. Caution. In a ringed space category $\mathcal{C}$ there will frequently be a notion of "sub-object". If A and B are objects of $\mathcal{C}$ and B is a "sub-object" of A then (in all reasonable cases) B will be a subset of A and the inclusion map $i : B \to A$ will be a morphism of $\mathcal{C}$, i.e., the restriction map, $f \mapsto f|B$ will be a homomorphism i^* of $\mathcal{C}(A)$ into $\mathcal{C}(B)$. However, in general i^* will not be surjective, so B will not be a ringed subspace of A (sub-object in the category of all ringed spaces). The problem of characterizing the pairs (A, B) such that B is a ringed subspace of A is of course an extension problem (when does each element of $\mathcal{C}(B)$ extend to an element of $\mathcal{C}(A)$) and is likely to be important, interesting and non-trivial. In the case of completely regular spaces (1.6.2) for example, the Tietze extension theorem says that a closed subspace B of a space A is a ringed subspace of A. Similarly if $\mathcal{C} = C^k$ (cf. 1.6.4) then for $k = 1, 2, \ldots, \infty$ it is a standard, rather elementary result that if B is a closed, regularly embedded, C^k submanifold of A then any C^k real valued function on B can be extended to a C^k real valued function on A, so that B is a ringed subspace of A. If $k = \omega$ then it is a deep, remarkable (and not so well-known) fact that the same result is still true. Since we shall wish to refer to these facts in the sequel we shall state them carefully below with an indication of where to find the proofs.

1.6.7. Remark. Let M be a C^ω manifold and let $N \subseteq M$. We recall that N is a closed, regularly embedded C^ω submanifold of M if N is a closed subset of M and for each $p \in N$ there is a neighborhood $\mathcal{O}$ of p in M and a C^ω chart $\psi : \mathcal{O} \to \mathbb{R}^m$ (mapping $\mathcal{O}$ say onto $\{x \in \mathbb{R}^m \mid |x_i| < 1, i = 1, \ldots, m\}$) such that ψ maps $\mathcal{O} \cap N$ onto $\{x \in \mathbb{R}^m \mid |x_i| < 1, i = 1, \ldots, m, x_i = 0, i = n+1, \ldots, m\}$. It follows of course that N is itself a C^ω manifold (the C^ω atlas coming from charts of the form $\psi \mid (N \cap \mathcal{O})$ as above) and that the inclusion map is a proper C^ω embedding. Conversely the image of any proper, C^ω embedding of a C^ω manifold into M is easily seen (by the C^ω implicit function theorem) to be a closed, regularly embedded, C^ω submanifold of M. To appreciate the following results note that *a priori* it might seem possible that every C^ω real valued function on M which vanished on N was identically zero and that in general a C^ω real valued function on N could not be extended to a C^ω function on all of M. (In fact this might seem to be possible even if N consisted only of two points; that is, it is not immediately evident that on an arbitrary C^ω manifold M there exist *any* non-constant C^ω real valued functions!)

1.6.8. *Theorem.* Every C^ω manifold M admits an embedding as a closed, regularly embedded, C^ω submanifold of some $\mathbb{R}^n$.

Proof. This theorem was proved first for M compact by C. B. Morrey in [22]. The general (paracompact) case was proved by H. Grauert in [11].

■

1.6.9. *Theorem.* Let M be a C^ω manifold, considered as a ringed space with structure ring $C^\omega(M, \mathbb{R})$ and let N be a closed, regularly embedded C^ω

submanifold of M. Then N is a Z-closed ringed subspace of M. That is, the restriction map $C^{\omega}(M, \mathbb{R}) \to C^{\omega}(N, \mathbb{R})$ is surjective and for each point $p \in M-N$ there is an $f \in C^{\omega}(M, \mathbb{R})$ which vanishes on N but not at p (in fact, there is an $f \in C^{\omega}(M, \mathbb{R})$ such that $N = f^{-1}(0)$).

<u>Proof.</u> By 1.6.8 we can assume M is a closed, regularly embedded C^{ω} submanifold of $\mathbb{R}^n$, and hence so is N. But this means we can assume $M = \mathbb{R}^n$ (for if $g \in C^{\omega}(N, \mathbb{R})$ and $g = G|N$ where $G \in C^{\omega}(\mathbb{R}^n, \mathbb{R})$ then $G|M \in C^{\omega}(M, \mathbb{R})$ and $g = (G|M)|N$, and if $N = F^{-1}(0)$ with $F \in C^{\omega}(\mathbb{R}^n, \mathbb{R})$ then $N = f^{-1}(0)$ where $f = F|M \in C^{\omega}(M, \mathbb{R})$). Now when $M = \mathbb{R}^n$ this theorem is due to H. Cartan [5]. (See statements (1) and (2) of §7, p. 89, of Cartan's paper and also the final sentence in the definition on p. 88. Note that in statement (1) Cartan claims only that there are a finite number of $f_i \in C^{\omega}(\mathbb{R}^n, \mathbb{R})$ such that $N = \bigcap_i f_i^{-1}(0)$, but clearly if we take $f = \sum_i f_i^2$ then $N = f^{-1}(0)$ and $f \in C^{\omega}(\mathbb{R}^n, \mathbb{R})$). ∎

1.6.10. <u>Remark.</u> Let M be a C^{ω} manifold and let N be a closed, regularly embedded C^{ω} submanifold of M. Given $p \in M$ there exist C^{ω} real valued functions $f_1, \ldots, f_m$, defined and forming a C^{ω} coordinate system in some neighborhood $\mathcal{O}$ of p in M, such that $N \cap \mathcal{O} = \{x \in \mathcal{O} \mid f_i(x) = 0,\ i = n+1, \ldots, m\}$ (1.6.7). It will be important for later purposes to know that we can describe N locally in this way but with the added condition that the functions $f_1, \ldots, f_m$ are global, that is, are the restrictions to $\mathcal{O}$ of functions belonging to the structure ring $C^{\omega}(M, \mathbb{R})$.

This is a non-trivial result and before proving it we shall need an additional fact, which moreover is interesting in its own right, namely, that if N is connected then all points of N look the same, and in fact the way N is embedded in M looks the same from all points of N. To be precise we shall show that if $p, q \in N$ then there exists a C^ω diffeomorphism of M which maps N onto itself and maps p onto q. For this in turn we shall have to appeal to the following well-known result.

Lemma. Let X be a C^ω vector field in $\mathbb{R}^n$ and let $\{\varphi_t\}$ be the maximal analytic flow generated by X. If $\|X\|$ is bounded, then in fact $\{\varphi_t\}$ is a one parameter group of global C^ω diffeomorphism of $\mathbb{R}^n$.

Proof. Let $p \in \mathbb{R}^n$ and let $(a,b) \subseteq \mathbb{R}$ be the interval on which $\varphi_t(p)$ is defined, i.e., the map $\sigma : (a,b) \to \mathbb{R}^n$, $t \mapsto \varphi_t(p)$ is the maximum solution curve of X with $\varphi_0(p) = p$. We must show that $a = -\infty$ and $b = \infty$. Suppose for example $b < \infty$. Then if $B = \operatorname{Sup} \|X\|$, the length of $t \mapsto \varphi_t(p)$ for $0 \le t \le b$ is less than or equal to bB, so for all t in this interval $\varphi_t(p)$ is in the (compact) closed ball of radius bB about p and hence $\varphi_t(p)$ has a limit point q as $t \to b$. By the local existence theorem for solutions of ordinary differential equations there is an $\epsilon > 0$ such that any solution curve of X which is inside the ϵ ball about q at time t_0 can be extended to a solution curve of X at least on the interval $(t_0-\epsilon, t_0+\epsilon)$. If we choose a t_0 in $(b-\epsilon, b)$ such that $\varphi_{t_0}(p)$ is in the ϵ ball about q it follows that $t \to \varphi_t(p)$ can be extended to an integral curve of X on $(a, t_0+\epsilon)$. But since $b-\epsilon < t_0$, $t_0+\epsilon > b$ contradicting the maximality of σ. ∎

1.6.11. Lemma. Let M be a closed, regularly embedded C^{ω} submanifold of $\mathbb{R}^k$ and let N be a closed, regularly embedded C^{ω} submanifold of M. Then there is a C^{ω} map P of $\mathbb{R}^k$ into the vector space $L(\mathbb{R}^k, \mathbb{R}^k)$ of linear endomorphisms of $\mathbb{R}^k$ (identified, as usual, with k by k matrices) such that:

1) For $x \in M$ $\operatorname{im}(P(x)) \subseteq TM_x$.

2) For $x \in N$ $\operatorname{im}(P(x)) = TN_x$ and $P(x)$ maps TN_x onto itself.

3) $P(x)$ is norm decreasing for all $x \in \mathbb{R}^k$.

Proof. We note that if P satisfies 1) and 2) and we replace P by $(1/f)P$, where $f \in C^{\omega}(\mathbb{R}^n, \mathbb{R})$ is a nowhere vanishing function, then this new P still clearly satisfies 1) and 2). If for f we take $\sqrt{k}(1+\sum_{i,j} P_{ij}^2)$ then each entry of the new matrix will be less than $(1/\sqrt{k})$ in absolute value, so that 3) follows from Schwartz's inequality. Hence it suffices to construct P satisfying 1) and 2). If we can construct a C^{ω} map $Q : \mathbb{R}^k \to L(\mathbb{R}^k, \mathbb{R}^k)$ such that for $x \in M$, $Q(x)$ is orthogonal projection on TM_x, then in the same way we can construct $R : \mathbb{R}^k \to L(\mathbb{R}^k, \mathbb{R}^k)$ such that for $x \in N$, $R(x)$ is orthogonal projection on TN_x and then define $P = QR$ (i.e., $P_{ij} = \sum_{\ell} Q_{i\ell} R_{\ell j}$). But then it will suffice to prove that the map $Q : M \to L(\mathbb{R}^k, \mathbb{R}^k)$ which maps x to orthogonal projection on TM_x is analytic, for then appealing to 1.6.9, we can extend Q to a C^{ω} map $\mathbb{R}^k \to L(\mathbb{R}^k, \mathbb{R}^k)$ by extending each Q_{ij} to an element of $C^{\omega}(\mathbb{R}^k, \mathbb{R})$). But the analyticity of Q (that is, of each Q_{ij}) is easy. Given $p \in M$ choose local C^{ω} coordinates $y_1, \ldots, y_k$ in a neighborhood $\mathcal{O}$ of p in $\mathbb{R}^k$ such that $U = M \cap \mathcal{O} = \{g \in \mathcal{O} \mid y_{m+1}(q) = \ldots = y_k(q) = 0\}$.

Then $(\partial/\partial y_1), \ldots, (\partial/\partial y_k)$ are C^ω vector fields in $\mathcal{O}$ so if we orthonormalize them by the Gram-Schmidt process the resulting vector fields $v_1, \ldots, v_k$ will also be C^ω in $\mathcal{O}$, and hence these restrictions to U (a neighborhood of p in M) will be C^ω in the analytic structure of M. Now for $x \in U$ and $v \in \mathbb{R}^k$ we clearly have $Q(x)v = \sum_{\ell=1}^{m} \langle v, v_\ell(x)\rangle v_\ell(x)$. Thus writing $e_1, \ldots, e_k$ for the standard basis of $\mathbb{R}^k$ we have

$$Q_{ij}(x) = \langle Q(x)e_i, e_j\rangle = \sum_{\ell=1}^{m} \langle e_i, v_\ell(x)\rangle\langle v_\ell(x), e_j\rangle .$$

Since the v_ℓ are C^ω maps of U into $\mathbb{R}^k$, their components $\langle v_\ell, e_j\rangle = \langle e_j, v_\ell\rangle$ are C^ω maps of U into $\mathbb{R}$. It follows that Q_{ij} is a C^ω map of U into $\mathbb{R}$, i.e., Q is a C^ω map $U \to L(\mathbb{R}^k, \mathbb{R}^k)$. ■

1.6.12. <u>Theorem.</u> Let M be a C^ω manifold and let N be a closed regularly embedded C^ω submanifold of M. Let G denote the group of C^ω diffeomorphisms φ of M such that $\varphi(N) = N$ and let G_0 denote the subgroup of G generated by elements belonging to analytic one parameter subgroups of G. Then G_0 acts transitively on each component of N.

<u>Proof.</u> Since N is the disjoint union of its G_0 orbits, it will suffice to prove that each orbit is open in N. Let $p \in N$. We will prove that $G_0 p$ is open in N by constructing a C^ω map $\psi : TN_p \to N$ such that $\psi(0) = p$, $T\psi_0 : TN_p \to TN_p$ is an isomorphism, and $\psi(v) = g_v(p)$ for some $g_v \in G_0$.

Define a map $X : \mathbb{R}^k \times TN_p \to \mathbb{R}^k$ by $X(x, v) = P(x)v$. Here we have used 1.6.8 to embed M as a closed, regularly embedded C^ω submanifold of $\mathbb{R}^k$ and then chosen $P : \mathbb{R}^k \to L(\mathbb{R}^k, \mathbb{R}^k)$ satisfying 1.6.11. Clearly X

is a C^ω map of $\mathbb{R}^k \times TN_p$ into $\mathbb{R}^k$ and, for each $v \in TN_p$, $x \mapsto X(x,v)$ is a bounded C^ω vector field on $\mathbb{R}^k$ (by 3) of 1.6.11) which is tangent to M at each point of M and tangent to N at each point of N (by 1) and 2) of 1.6.11). Let $\{\varphi_t^v\}$ be the one-parameter group of C^ω diffeomorphisms of $\mathbb{R}^k$ generated by this vector field (see the lemma of 1.6.10). By the above tangency properties it is clear that $\varphi_t^v(M) = M$ and $\varphi_t^v(N) = N$. By the classical fact that if a differential equation depends C^ω on parameters then the solutions depend analytically on these parameters, it follows that $(t,x,v) \mapsto \varphi_t^v(x)$ is actually a C^ω map $\mathbb{R} \times \mathbb{R}^k \times TN_p \to \mathbb{R}^k$. Since $X(x,v)$ is clearly linear in v we note that $\varphi_t^v(x) = \varphi_1^{tv}(x)$. Thus if we define a C^ω map $\Phi : \mathbb{R}^k \times TN_p \to \mathbb{R}^k$ by $(x,v) \mapsto \varphi_1^v(x)$ we have $\varphi_t^v(x) = \Phi(x, tv)$. Let $\Psi : M \times TN_p \to M$ denote the restriction of Φ (recall that $\Psi(x,v) = \Phi(x,v) = \varphi_1^v(x) \in M$ if $x \in M$). Then $(t,x) \mapsto \Psi(x,tv)$ is for each $v \in TN_p$ a one-parameter group of G (since $\Psi(x,v) = \varphi_1^v(x) \in N$ if $x \in N$) and hence $x \mapsto \Psi(x,v)$ is for each $v \in TN_p$ an element of G_0. Thus it will suffice to show that the map $\psi : TN_p \to N$ defined by $\psi(v) = \Psi(p,v)$ covers a neighborhood of p in N. Since $X(x,0) = P(x)0 = 0$ we have $\psi(0) = p$ and hence by the inverse function theorem it will suffice to show that $T\psi_0$, the differential of ψ at 0, maps $T(TN_p)_0 = TN_p$ onto itself. But $T\psi_0(v)$ is just the tangent vector at $t = 0$ to the curve $t \mapsto \psi(tv) = \varphi_t^v(p)$, which by definition is the solution curve of the vector field $X(\cdot\,, v)$ starting at p. Thus $T\psi_0(v) = X(p,v) = P(p)v$, i.e., $T\psi_0 = P(p)$, and by 2) of 1.6.11 $P(p)$ maps TN_p onto itself. ∎

1.6.13. Remark. Let $x_1, \ldots, x_k$ denote the standard linear coordinate system in $\mathbb{R}^k$. Let $0 \leq \rho < k$ and for each $(\rho+1)$-tuple of integers μ with $1 \leq \mu_1 < \ldots < \mu_{\rho+1} \leq k$ let dx_μ denote the $(\rho+1)$-form $dx_{\mu_1} \wedge \ldots \wedge dx_{\mu_{\rho+1}}$ on $\mathbb{R}^k$ (we will regard $(\rho+1)$-forms on $\mathbb{R}^k$ as maps of $\mathbb{R}^k$ into $\Lambda^{\rho+1}(\mathbb{R}^{k*})$, so the dx_μ are actually constant maps onto the standard basis for $\Lambda^{\rho+1}(\mathbb{R}^{k*})$).

Now let $y_1, \ldots, y_k \in C^\omega(\mathbb{R}^k, \mathbb{R})$ and suppose that they form a C^ω coordinate system in some open set $\mathcal{O}$ of $\mathbb{R}^k$. Then $dy_\mu = dy_{\mu_1} \wedge \ldots \wedge dy_{\mu_{\rho+1}}$ are C^ω $(\rho+1)$-forms in $\mathbb{R}^k$ and at each point of $\mathcal{O}$ the dy_μ form a basis for $\Lambda^{\rho+1}(\mathbb{R}^{k*})$. It follows that:

Lemma a. There are uniquely determined elements $J^\mu_{\bar{\mu}}$ of $C^\omega(\mathcal{O}, \mathbb{R})$ (one for each pair $\mu, \bar{\mu}$ of $(\rho+1)$-tuples as above) such that $dx_\mu = \sum_{\bar{\mu}} J^\mu_{\bar{\mu}} dy_{\bar{\mu}}$ in $\mathcal{O}$. (In fact $J^\mu_{\bar{\mu}}$ is the Jacobian determinant $(\partial x_\mu / \partial y_{\bar{\mu}}) = \partial(x_{\mu_1}, \ldots, x_{\mu_{\rho+1}})/\partial(y_{\bar{\mu}_1}, \ldots, y_{\bar{\mu}_{\rho+1}})$.)

Definition. For each $g \in C^\omega(\mathbb{R}^k, \mathbb{R})$ we define elements $\Phi_\mu(g)$ of $C^\omega(\mathbb{R}^k, \mathbb{R})$ by $dy_1 \wedge \ldots \wedge dy_\rho \wedge dg = \sum_\mu \Phi_\mu(g) dx_\mu$.

Lemma b. All the functions $\Phi_\mu(g)$ vanish at any point where $dy_1, \ldots, dy_\rho$, dg are linearly dependent.

Proof. Trivial.

<u>Lemma</u> c. If $\rho < j \leq k$ then in $\mathcal{O}$ we have $\partial g/\partial y_j = \sum_\mu A_\mu \Phi_\mu(g)$ where $A_\mu \in C^\omega(\mathcal{O}, \mathbb{R})$ is independent of g. In fact $A_\mu = \partial(x_{\mu_1}, \ldots, x_{\mu_{\rho+1}})/\partial(y_1, \ldots, y_\rho, y_j))$.

<u>Proof.</u> Since $dy_1 \wedge \ldots \wedge dy_\rho \wedge dy_j = 0$ for $j \leq \rho$, and since in $\mathcal{O}$ we have $dg = \sum_{j=1}^{k} (\partial g/\partial y_j) dy_j$ we have in $\mathcal{O}$ the identity

$$\sum_{j=\rho+1}^{k} (\partial g/\partial y_j) dy_1 \wedge \ldots \wedge dy_\rho \wedge dy_j = dy_1 \wedge \ldots \wedge dy_\rho \wedge dg$$

$$= \sum_\mu \Phi_\mu(g) dx_\mu$$

$$= \sum_{\mu, \overline{\mu}} \Phi_\mu(g) J^\mu_{\overline{\mu}} dy_{\overline{\mu}} .$$

But since the $dy_{\overline{\mu}}$ form a basis for $\Lambda^{\rho+1}(\mathbb{R}^{k*})$ at each point of $\mathcal{O}$ we can "equate coefficients" in the above identity. On the left the coefficient of $dy_1 \wedge \ldots \wedge dy_\rho \wedge dy_j$ is $(\partial g/\partial y_j)$ while on the right is $\sum_\mu \Phi_\mu(g) J^\mu_{\overline{\mu}}$ where $\overline{\mu} = (1, 2, \ldots, \rho, j)$. ∎

1.6.14. <u>Theorem.</u> Let M be a C^ω manifold and let N be a closed, regularly embedded C^ω submanifold of M. Let $I = \{f \in C^\omega(M, \mathbb{R}) \mid (f|N) = 0\}$ and for $p \in N$ let V_p denote the subspace of T^*M_p spanned by $\{df_p | f \in I\}$. Then:

1) The dimension ρ of V_p is independent of p and in fact $\rho = \dim M - \dim N$.

2) Let $p_0 \in N$ and let $y_1, \ldots, y_\rho \in I$ be such that $(dy_1)_{p_0}, \ldots, (dy_\rho)_{p_0}$ are a basis for V_{p_0}. Then there exist $y_{\rho+1}, \ldots, y_m$ in $C^\omega(M, \mathbb{R})$ such that

$(dy_1)_{p_0}, \ldots, (dy_m)_{p_0}$ is a basis for $T^*M_{p_0}$ and hence $y_1, \ldots, y_m$ is a C^ω coordinate system for M near p_0; i.e., if $\epsilon > 0$ is sufficiently small then there is a neighborhood $\mathcal{O}$ of p_0 in M such that $p \mapsto (y_1(p), \ldots, y_m(p))$ maps $\mathcal{O}$ C^ω diffeomorphically onto $\{x \in \mathbb{R}^m \mid |x_i - y_i(p_0)| < \epsilon\}$. Moreover, if ϵ is sufficiently small then $\mathcal{O} \cap N$ is equal to $S = \{p \in \mathcal{O} \mid y_1(p) = \ldots = y_\rho(p) = 0\}$.

Proof. We first consider the special case that $M = \mathbb{R}^k$ and N is connected. It follows from 1.6.12 that $\dim V_p$ is a constant ρ (for if $\varphi \in G$, then clearly $f \mapsto f \circ \varphi^{-1}$ is an automorphism of $C^\omega(M, \mathbb{R})$ which preserves I and so induces an isomorphism of V_p onto $V_{\varphi(p)}$). It will follow trivially from 2) that $\rho = \dim M - \dim N$.

The existence of $y_{\rho+1}, \ldots, y_k$ in $C^\omega(\mathbb{R}^k, \mathbb{R})$ such that $(dy_1)_{p_0}, \ldots, (dy_k)_{p_0}$ is a basis for $T^*\mathbb{R}^k_{p_0} = \mathbb{R}^{k*}$ is trivial, and we can in fact even choose them to be linear. Since $y_1, \ldots, y_\rho$ belong to I it is clear that $\mathcal{O} \cap N \subseteq S$ so it remains only to prove the reverse inclusion, $S \subseteq \mathcal{O} \cap N$. Now by 1.6.9 $N = \{p \in \mathbb{R}^k \mid g(p) = 0 \text{ for all } g \in I\}$ so it will suffice to show that if $g \in I$ then $g|S = 0$. The following beautiful proof of this fact is due to H. Whitney [32]. Using the notation of 1.6.13 we see from Lemma b of that section that $g \in I$ implies $\Phi_\mu(g) \in I$ for all μ (for since $\dim V_p = \rho$, and $f_1, \ldots, f_\rho$, g all belong to I, at every $p \in N$, $(df_1)_p, \ldots, (df_\rho)_p$ and $(dg)_p$ are linearly dependent). Then by Lemma c it follows from the rule for differentiating a product and a trivial induction that any partial derivative $(\partial^r_g / \partial y_{j_1}, \ldots, \partial y_{j_r})$, with all the $j_i > \rho$, can be written in $\mathcal{O}$ as a sum of terms of the form Ah where $A \in C^\omega(\mathcal{O}, \mathbb{R})$ and $h \in I$, and in particular all such partial derivatives vanish at points of $\mathcal{O} \cap N$. Now $p_0 \in S$ is in $\mathcal{O} \cap N$ so all the partial derivatives of g with respect to $y_{\rho+1}, \ldots, y_k$ vanish at p_0. But S is a connected C^ω submanifold of $\mathcal{O}$, $g|S \in C^\omega(S, \mathbb{R})$ and $y_{\rho+1}, \ldots, y_k$ is a C^ω coordinate system for S at p_0. Since the Taylor series for $g|S$ vanishes at p_0 it follows from the principle of analytic continuation that $g|S = 0$, which completes the proof of the special case.

Next suppose N is not necessarily connected and let N_0 be a component of N. Then N_0 is a closed regularly embedded C^ω submanifold of M so we can apply the preceding result to N_0. Let $I_0 = \{f \in C^\omega(M, \mathbb{R}) \mid (f|N_0) = 0\}$ and note that $I \subseteq I_0$. The theorem for N will follow trivially from the theorem for N_0 if we can show that for $p \in N_0$, $\{df_p \mid f \in I\} = \{df_p \mid f \in I_0\}$. Now the function k which is one on N_0 and zero on the other components of N is clearly in $C^\omega(N, \mathbb{R})$ so by 1.6.9 it equals $H|N$ for some $H \in C^\omega(M, \mathbb{R})$. If $f \in I_0$ then $Hf \in I$ and

$$d(Hf)_p = H(p)df_p + f(p)dH_p = df_p$$

which proves that $\{df_p \mid f \in I_0\} \subseteq \{df_p \mid f \in I\}$ and the reverse inclusion follows from $I \subseteq I_0$.

Finally, we must consider the case of general M, not necessarily equal to $\mathbb{R}^k$. In any case we can by 1.6.8 assume that M is a closed, regularly embedded C^ω submanifold of $\mathbb{R}^k$. Let $\tilde{I} = \{F \in C^\omega(\mathbb{R}^k, \mathbb{R}) \mid (F|N) = 0\}$ and note that by 1.6.9 the restriction map $F \mapsto F|M$ is a homomorphism of $\tilde{I}$ onto

I. Choose $Y_1, \ldots, Y_\rho \in \widetilde{I}$ such that $y_i = Y_i|M$. Note that since $(dy_i)_p = (dY_i)_p|TM_p$ are linearly independent so, a fortiori, are the $(dY_i)_p$; hence we can choose $Y_{\rho+1}, \ldots, Y_\sigma \in \widetilde{I}$ so that $(dY_1)_p, \ldots, (dY_\sigma)_p$ is a basis for the subspace $\{dF_p | F \in \widetilde{I}\}$ of $\mathbb{R}^{k*}$. Note that $(dY_{\rho+1})_p|TM_p, \ldots, (dY_\sigma)_p|TM$ depend linearly on $(dY_1)_p|TM_p, \ldots, (dY_\rho)_p|TM$. By the special case of the theorem we can choose $Y_{\sigma+1}, \ldots, Y_k \in C^\omega(\mathbb{R}^k, \mathbb{R})$ forming a local coordinate system for $\mathbb{R}^k$ in a neighborhood $\widetilde{\mathcal{O}}$ of p in $\mathbb{R}^k$ and $\widetilde{\mathcal{O}} \cap N = \{x \in \widetilde{\mathcal{O}} | Y_1(x) = \ldots = Y_\sigma(x) = 0\}$, and hence $Y_{\sigma+1}, \ldots, Y_k$ restricted to N are local coordinates for N at p, and in particular $\dim N = (k-\sigma)$. Putting $y_i = Y_i|M$ it will then suffice to show that $y_1, \ldots, y_\rho, y_{\sigma+1}, \ldots, y_k$ is a local coordinate system for M near p, so that in particular $\dim M = \rho+(\sigma-k)$ and hence $\rho = \dim M - \dim N$. For then putting $\mathcal{O} = \widetilde{\mathcal{O}} \cap M$, with $\widetilde{\mathcal{O}}$ sufficiently small, it is clear that $\mathcal{O} \cap N \subseteq \{x \in \mathcal{O} | y_1(x) = \ldots = y_\rho(x) = 0\}$ since $y_1, \ldots, y_\rho$ belong to I, and the reverse inclusion then follows by dimensionality considerations. Thus we must show that $(dy_1)_p, \ldots, (dy_\rho)_p, (dy_{\sigma+1})_p, \ldots, (dy_k)_p$ are a basis for T^*M_p. Since $(dy_i)_p$, $i = 1, \ldots, k$, clearly span T^*M_p (because the $(dY_i)_p$ span $\mathbb{R}^{k*}$) and $(dy_{\rho+1})_p, \ldots, (dy_\sigma)_p$ depend linearly on $(dy_1)_p, \ldots, (dy_\rho)_p$ what we must show is that $(dy_1)_p, \ldots, (dy_\rho)_p, (dy_{\sigma+1})_p, \ldots, (dy_k)_p$ are linearly independent. Since $(dy_1)_p, \ldots, (dy_\rho)_p$ is a basis for V_p and $(dy_{\sigma+1})_p, \ldots, (dy_k)_p$ restricted to TN_p is a basis for TN_p it suffices then to note that any element of V_p when restricted to TN_p is zero. ∎

1.7. The Irreducible Components of a Space.

1.7.1. Proposition and Definition. Let X be a non-empty topological space. The following three properties are equivalent and if X satisfies any one and hence all of them, it is called irreducible.

1) X is not the union of two proper closed subsets.

2) Each non-empty open set of X is dense in X.

3) Every open set of X is connected.

Proof. ((1) $\Rightarrow$ (2)). Given $\mathcal{O}$ open in X and non-empty, note that $\overline{\mathcal{O}}$ and $X-\mathcal{O}$ are closed and $X-\mathcal{O}$ is a proper subset of X. It follows from $X = \overline{\mathcal{O}} \cup (X-\mathcal{O})$ that $\overline{\mathcal{O}} = X$.

((2) $\Rightarrow$ (3)). If the open set $\mathcal{O}$ is the union of two disjoint subsets G_1 and G_2, open in $\mathcal{O}$ and hence in X, then either $\mathcal{O}$ is empty (hence connected) or else one of G_1 or G_2, say G_1, is not empty, hence dense in X. Since G_2 is open in X and does not meet G_1, $G_2 = \phi$ so $\mathcal{O}$ is connected.

((3) $\Rightarrow$ (1)). Let X be the union of closed sets F_1 and F_2. Then $X-F_1$ and $X-F_2$ are disjoint open sets. Since their union is connected, one of them is empty; i.e., not both F_1 and F_2 are proper. ■

1.7.2. Exercise. Show that an irreducible Hausdorff space contains only one point.

1.7.3. Proposition. If Y is a dense subspace of X then Y is irreducible if and only if X is irreducible.

Proof. Suppose Y is irreducible and let $\mathcal{O}$ be open in X and non-empty. Since Y is dense in X, $\mathcal{O} \cap Y$ is non-empty and of course open in Y; hence $\mathcal{O} \cap Y$ is dense in Y, hence in $\overline{Y} = X$. A fortiori $\mathcal{O}$ is also dense in X.

Suppose X is irreducible and let $Y = F_1 \cup F_2$ where F_1 and F_2 are closed in Y. Then $X = \overline{Y} = \overline{F}_1 \cup \overline{F}_2$, so one of $\overline{F}_1$ and $\overline{F}_2$, say $\overline{F}_1$, is equal to X, say $\overline{F}_1 = X$. In other words F_1 is dense in X and a fortiori F_1 is dense in Y. But F_1 is closed in Y and hence $F_1 = Y$ and Y is irreducible.

■

1.7.4. Corollary. If Y is a subspace of X then Y is irreducible if and only if $\overline{Y}$ is irreducible.

1.7.5. Corollary. The closure of a point is irreducible.

1.7.6. Proposition. The union of a chain of irreducible subspaces of X is irreducible.

Proof. Suppose $X \supseteq Y = \bigcup_\alpha Y_\alpha$ with each Y_α irreducible and for all α, β either $Y_\alpha \subseteq Y_\beta$ or $Y_\beta \subseteq Y_\alpha$. Let $X = F_1 \cup F_2$ with F_1 and F_2 closed in X. Since $Y_\alpha = (Y_\alpha \cap F_1) \cup (Y_\alpha \cap F_2)$ it follows that each Y_α is included either in F_1 or F_2. Suppose $Y_{\alpha_0} \not\subseteq F_1$ for some α_0. Then for all α with $Y_{\alpha_0} \subseteq Y_\alpha$ we have $Y_\alpha \not\subseteq F_1$ so $Y_\alpha \subseteq F_2$. But Y is the union of the Y_α with $Y_{\alpha_0} \subseteq Y_\alpha$; hence $Y \subseteq F_2$. ■

1.7.7. Definition. A maximal irreducible subspace of a topological space X is called an irreducible component of X.

1.7.8. Proposition. Every irreducible subspace of a space X is included in some irreducible component of X.

Proof. Immediate from 1.7.6 and Zorn's Lemma. ■

1.7.9. Corollary. Every topological space X is the union of its irreducible components.

Proof. Each point of X is an irreducible subspace of X and so by 1.7.8 is contained in some irreducible component of X.

1.7.10. Proposition. An irreducible component of X is a closed, connected subspace of X.

Proof. If Y is an irreducible component of X then since $Y \subseteq \overline{Y}$ it follows from 1.7.4 that $Y = \overline{Y}$. That Y is connected is immediate from (3) of 1.7.1. ■

1.7.11. Proposition. If X is the union of closed subspaces $X_1, \ldots, X_n$, then each irreducible subspace of X is included in one of the X_i.

Proof. If $n = 1$ there is nothing to prove. If $n = 2$ the proposition follows from (1) of 1.7.1. Since $X_1 \cup \ldots \cup X_{n-1}$ is closed the general case follows by a trivial induction. ■

1.7.12. Corollary. If a topological space X is a finite union of closed, irreducible subspaces $X_1, \ldots, X_n$, then each irreducible component of X is one of the X_i.

Proof. Trivial. ■

1.7.13. Theorem. Let X be a topological space which can be expressed as a finite union of closed, irreducible subspaces:

$$X = X_1 \cup \ldots \cup X_n.$$

Then it is possible to do this in exactly one way (up to order) such that none of the X_i are included in any of the others. The subspaces of X that occur in such an "irredundant" representation of X as the union of closed, irreducible subspaces are exactly the irreducible components of X. Moreover, none of them is included in the union of any others.

Proof. The existence of an irredundant representation of X is clear, since as long as there is an X_i occurring which is included in some other X_j we can delete it. That every irreducible component of X is one of the remaining X_i is clear from 1.7.12. It now follows that each X_i is an irreducible component. For by 1.7.8, X_i is in any case included in some irreducible component of X, i.e., in some X_j, and $j \neq i$ would contradict irredundancy. The final statement of the theorem is clear from 1.7.11. ■

1.7.14. Proposition. Suppose a space X has only finitely many irreducible components $X_1, \ldots, X_n$ and that they are disjoint. Then the X_i

are also the connected components of X.

Proof. Since the X_i are connected and disjoint it will suffice to show that each is open in X, i.e., that $(X-X_i) = \bigcup_{j\neq i} X_j$ is closed in X. But since each X_j is closed, this is clear. ■

1.7.15. Theorem. Let X be a topological space having only finitely many irreducible components, $X_1, \ldots, X_n$. Then:

1) If S is dense in X then $S_i = S \cap X_i$ is dense in X_i and $S_1, \ldots, S_n$ are the irreducible components of S.
2) If U is open in X and meets each X_i then U is dense in X.
3) $X_i' = X_i - \bigcup_{j\neq i} X_j$ is open in X.
4) $U_0 = \bigcup_i X_i'$ is an open dense subset of X whose irreducible and connected components are the X_i'.

Proof. By 1.7.13, X_i is not included in $\bigcup_{j\neq i} X_j$ so a fortiori X_i is not included in $\bigcup_{j\neq i} \overline{S}_j$. On the other hand, $X_i \subseteq X = \overline{S} = \overline{S}_i \cup \bigcup_{j\neq i} \overline{S}_j$. Since X_i is irreducible it follows that $X_i \subseteq \overline{S}_i$ so S_i is dense in X_i. By 1.7.4, S_i is irreducible and clearly S_i is closed in S. Since $S_i \subseteq S_j$ would imply $X_i = \overline{S}_i \subseteq \overline{S}_j = X_j$ it follows from 1.7.13 that the S_i are the irreducible components of S.

The second statement is clear from (2) of 1.7.1.

Since X_j is closed in X and $X_i' = X - \bigcup_{j\neq i} X_j$ it is clear that X_i'

is open in X. Moreover, $X_i' \neq \phi$ since otherwise we would have $X_i \subseteq \bigcup_{j \neq i} X_j$.
Then conclusions 3) and 4) now follow from 1) and 2) and 1.7.14. ■

1.7.16. Definition. Let $(S, \mathcal{Q})$ be a ringed space. We say that S is irreducible if it is irreducible in its Z-topology and we say that S has the unique continuation property if whenever f and g in $\mathcal{Q}$ agree on a non-empty Z-open subset of S they are equal (or equivalently if $f \in \mathcal{Q}$ vanishes on a non-empty Z-open subset of S then $f = 0$). If X is a subset of S then we say that X is irreducible or has the unique continuation property if it has these properties when regarded as a ringed subspace of $(S, \mathcal{Q})$.

1.7.17. Remark. Since the inclusion $X \hookrightarrow S$ is a homeomorphism into with respect to the Z-topologies (cf., 1.5.7) X is irreducible if and only if it is irreducible when considered as a subspace of S with the Z-topology. Similarly X has the unique continuation property if and only if whenever f and g in $\mathcal{Q}$ agree on a non-empty relatively Z-open subset of X they agree on X.

1.7.18. Theorem. If X is a subset of the ringed space $(S, \mathcal{Q})$ then the following are equivalent:

1) X has the unique continuation property.

2) X is irreducible.

3) $I(X) = \{f \in \mathcal{Q} \mid (f(X) = 0\}$ is a prime ideal of $\mathcal{Q}$.

4) $\mathcal{Q}(X) = \{(f(X) \mid f \in \mathcal{Q}\}$, the structure ring of X, is an integral domain.

In particular S itself is irreducible if and only if $\mathcal{Q}$ is an integral domain.

Proof. ((1) <=> (2)). The Z-closure of a subset A of X is the set of x in X such that every $f \in \mathcal{A}$ vanishing on A also vanishes at x. Thus the unique continuation property for X means that every non-empty Z-open subset of X is dense, i.e., that X is irreducible in its Z-topology.

((3) <=> (4)). $\mathcal{A}(X) \approx \mathcal{A}/I(X)$.

((2) ⇒ (3)). Given $fg \in I(X)$ let $Z(f) = \{x \in X \mid f(x) = 0\}$ and $Z(g) = \{x \in X \mid g(x) = 0\}$ so that $X = Z(f) \cup Z(g)$. Since Z(f) and Z(g) are closed, one of them is all of X; i.e., one of f and g vanishes on X; i.e., one of f and g belong to I(X).

((3) ⇒ (2)). Suppose X is the union of proper Z-closed subsets F_1 and F_2. Then there exists $f_i \in \mathcal{A}$ such that $f_i | F_i = 0$ but f_i does not vanish identically on X, so $f_1 f_2 \in I(X)$ but neither f_1 nor f_2 is in I(X). Hence ~(2) implies ~(3). ■

1.7.19. Corollary. If $(S, \mathcal{A})$ is a ringed space, then $X \to I(X)$ is an inclusion reversing bijective correspondence between the collection of irreducible Z-closed subsets of X and a certain collection of prime strict radical ideals of $\mathcal{A}$ (namely, those I whose variety $V = V(I) \subseteq \hat{\mathcal{A}}$ satisfies $V \cap S$ is Z-dense in V).

Proof. Immediate from 1.5.30. ■

1.7.20. Corollary. If $(S, \mathcal{A})$ is a complete ringed space, then $X \mapsto I(X)$ is a bijective correspondence between the irreducible components of S (relative to its Z-topology) and the minimal, prime, strict radical ideals of $\mathcal{A}$.

Proof. An irreducible component of S is just a maximal (automatically Z-closed) irreducible subspace so this follows directly from 1.7.19.

∎

1.7.21. Theorem. If $(S, \mathcal{A})$ is a ringed space and $\mathcal{A}$ has only finitely many minimal prime ideals, then S has only finitely many irreducible components $X_1, \ldots, X_n$ and $I(X_1), \ldots, I(X_n)$ are these minimal primes.

Proof. If S is complete this is immediate from 1.7.20 and 1.3.17. Since S is dense in $\hat{\mathcal{A}}$, it follows from 1.7.15 and 1.7.19 that the same holds even if S is not complete.

∎

1.7.22. Proposition. The product $S_1 \times S_2$ of ringed spaces S_1 and S_2 over K is irreducible if and only if both S_1 and S_2 are irreducible.

Proof. If S_1 is reducible, say $S_1 = A \cup B$ where A and B are proper Z-closed subsets of S_1, then $S_1 \times S_2 = (A \times S_2) \cup (B \times S_2)$ which are proper Z-closed subsets of $S_1 \times S_2$ (because $\prod_1 : S_1 \times S_2 \to S_1$ is a ringed space morphism and so Z-continuous, cf. 1.5.6 and 1.5.23). Thus $S_1 \times S_2$ is reducible.

Conversely, suppose S_1 and S_2 are irreducible and let $S_1 \times S_2 = A \cup B$ where A and B are Z-closed subsets of $S_1 \times S_2$. Given $s_1 \in S_1$ recall that $\{s_1\} \times S_2$ is a subspace of $S_1 \times S_2$ isomorphic to S_2 (cf. 1.5.23) and so irreducible; hence either $\{s_1\} \times S_2 \subseteq A$ or $\{s_1\} \times S_2 \subseteq B$. Let $S_1^A = \{s_1 \in S_1 \mid \{s_1\} \times S_2 \subseteq A\}$. We note that S_1^A is closed in S_1. (In fact, it is the intersection (over $s_2 \in S_2$) of the sets $\{s_1 \in S_1 | (s_1, s_2) \in A\}$ and the latter is the inverse image of A under the ringed space morphism $i_{s_2} : S_1 \to S_1 \times S_2$, $s_1 \mapsto (s_1, s_2)$ (cf. 1.5.23).) Similarly $S_1^B = \{s_1 \in S_1 | \{s_1\} \times S_2 \subseteq B\}$ is also a Z-closed subset of S_1. Since we know $S_1 = S_1^A \cup S_1^B$ and S_1 is irreducible either $S_1^A = S_1$ or $S_1^B = S_1$. But in the first case $S_1 \times S_2 = A$ and in the second $S_1 \times S_2 = B$, so not both A and B can be proper subsets of $S_1 \times S_2$ and $S_1 \times S_2$ is irreducible. ∎

1.8. Noetherian Spaces.

1.8.1. Definition. A topological space X is called Noetherian if it satisfies any one and hence all of the following three equivalent properties:

1) Every open set in X is compact.

2) The open subsets of X satisfy the ascending chain condition.

3) The closed subsets of X satisfy the descending chain condition.

1.8.2. Remark. The equivalence of 2) and 3) is immediate from De Morgan's laws. To see that 1) implies 2) let $\mathcal{O}_1 \subseteq \mathcal{O}_2 \subseteq \ldots$ be an ascending chain of open sets and let $\mathcal{O} = \bigcup_j \mathcal{O}_j$. Since $\mathcal{O}$ is compact it follows that $\mathcal{O} = \mathcal{O}_j$ for some j and hence that $\mathcal{O}_i = \mathcal{O}_j$ for $j > i$. To see that 2) implies 1), given $\mathcal{O}$ open and $\mathcal{O} = \bigcup_\alpha \mathcal{O}_\alpha$ with the $\mathcal{O}_\alpha$ open, pick $\mathcal{O}_{\alpha_1}$ and, as long as $\mathcal{O}_{\alpha_1} \cup \ldots \cup \mathcal{O}_{\alpha_n} \neq \mathcal{O}$ choose $\mathcal{O}_{\alpha_{n+1}}$ not included in the union of $\mathcal{O}_{\alpha_1}, \ldots, \mathcal{O}_{\alpha_n}$. Then $\mathcal{O}_{\alpha_1}, \mathcal{O}_{\alpha_1} \cup \mathcal{O}_{\alpha_2}, \ldots$ is an ascending chain of open sets, so must eventually become constant, i.e.; eventually $\mathcal{O}_{\alpha_1} \cup \ldots \cup \mathcal{O}_{\alpha_n} = \mathcal{O}$. ∎

1.8.3. Theorem. A Noetherian space X has only finitely many irreducible components. Equivalently (cf. 1.7.13) X admits an irredundant representation as the finite union of closed irreducible subspaces.

Proof. Assume X had infintely many irreducible components. We will show how to construct inductively an infinite strictly decreasing sequence $X \supset A_1 \supset \ldots \supset A_n \supset \ldots$ of closed subsets of X.

We know by 1.7.12 that X is not the union of a finite number of closed irreducible subspaces. In particular X is not itself irreducible, so $X = X_1 \cup A_1$ where X_1 and A_1 are closed, non-empty proper subsets of X. At least one of these, say A_1, cannot be written as a finite union of closed irreducible subspaces. Then $A_1 = X_2 \cup A_2$ where X_2 and A_2 are closed, non-empty, proper subsets of A_1. We leave the inductive step to the reader.

■

1.8.4. Remark. Recall that a commutative algebra $\mathcal{A}$ over K is called Noetherian if the ideals of $\mathcal{A}$ satisfy the ascending chain condition, or equivalently if every ideal of $\mathcal{A}$ is finitely generated (as an $\mathcal{A}$ module). Obviously K itself is Noetherian since its only ideals are (0) and (1) = K. According to the Hilbert basis theorem, if $\mathcal{A}$ is Noetherian, then so is the polynomial ring $\mathcal{A}[X]$, so inductively it follows that $K[X_1, \ldots, X_n]$ is Noetherian for all n. If $\mathcal{A}$ is Noetherian and $0 \to \mathcal{I} \to \mathcal{A} \to B \to 0$ is exact, then it is trivial that B is also Noetherian, for (identifying B with $\mathcal{A}/\mathcal{I}$) the ideals of B are of the form $J/\mathcal{I}$ where J is an ideal of $\mathcal{A}$ which includes $\mathcal{I}$. It follows that any commutative algebra B over K which is finitely generated (as an algebra) is Noetherian. For if $(\xi_1, \ldots, \xi_n)$ are generators for B then $P(X_1, \ldots, X_n) \mapsto P(\xi_1, \ldots, \xi_n)$ defines an epimorphism of $K[X_1, \ldots, X_n]$ onto B.

1.8.5. Definition. A ringed space $(S, \mathcal{A})$ is called Noetherian if the structure ring $\mathcal{A}$ is Noetherian.

1.8.6. Theorem. Let $(S, \mathcal{A})$ be a Noetherian ringed space. Then

S is Noetherian in its Z-topology, and if $X_1, \ldots, X_n$ are the irreducible components of S then $p_i = I(X_i)$ are the minimal prime ideals of $\mathcal{Q}$ and $X_i = \{x \in S \mid f(x) = 0 \text{ for all } f \in p_i\}$.

Proof. A strictly descending sequence of Z-closed sets $X_1 \supset X_2 \supset \ldots \supset X_n \supset \ldots$ would give a strictly ascending sequence $I(X_1) \subset \ldots \subset I(X_n) \subset \ldots$ of strict radical ideals of $\mathcal{Q}$ by 1.5.30. This shows that S is Noetherian in its Z-topology and then 1.8.3, 1.7.20 and 1.7.21 complete the proof. ∎

1.9. Tangent and Cotangent Spaces.

1.9.1. Definition. Let $(S,\mathcal{Q})$ be a ringed space over K and $s \in S$. A tangent vector to S at s is a linear (over K) map $D : \mathcal{Q} \to K$ such that:

$$D(fg) = (Df)(g(s)) + (f(s))(Dg).$$

The vector space (over K) of all tangent vectors to S at s is called the tangent space to S at s and is denoted by $T(S)_s$. If $\varphi : (S,\mathcal{Q}) \to (U,B)$ is a ringed space morphism we define a linear map $(T\varphi)_s : T(S)_s \to T(U)_{\varphi(x)}$ by $(T\varphi)_s(D) = D \circ \varphi^*$, i.e., for $f \in B$, $(T\varphi)_s(D)f = D(f \circ \varphi)$.

1.9.2. Proposition. Let $\varphi : (S,\mathcal{Q}) \to (U,B)$ be a ringed space morphism such that $\varphi^* : B \to \mathcal{Q}$ is surjective. Then $(T\varphi)_s : (TS)_x \to (TU)_{\varphi(s)}$ is injective for all $s \in S$. In particular, if S is a ringed subspace of U and φ is the inclusion map (so φ^* is restriction) we may regard $(T\varphi)_s$ as an identification of $T(S)_s$ with a subspace of $(TU)_{\varphi(s)}$.

Proof. If $(T\varphi)_s(D) = 0$ then $D(f \circ \varphi) = 0$ for all $f \in B$. But by assumption every $g \in \mathcal{Q}$ is of the form $\varphi^*(f) = f \circ \varphi$ for some $f \in B$, hence $D = 0$. ■

1.9.3. Proposition. If $D \in T(S)_s$ then D vanishes on elements of K and on m^2, where m denotes the maximal ideal in $\mathcal{Q}$ consisting of functions vanishing at s.

Proof. $D(1) = D(1\cdot 1) = 1(s)D(1) + D(1)1(s)$

$$= D(1)+D(1)$$

so $D(1) = 0$ and hence $D(c) = D(c1) = cD(1) = 0$ for all $c \in K$. If $f, g \in m$, i.e., if $f(s) = g(s) = 0$, then $D(fg) = (Df)(g(s))+f(s)(Dg) = 0$. Since elements of m^2 are finite sums of such products, D vanishes on m^2. ■

1.9.4. Definition. Let $(S, \mathcal{a})$ be a ringed space over K, $s \in S$ and let m denote the maximal ideal in $\mathcal{a}$ of functions vanishing at s. Then we define $T^*(S)_s$, the cotangent space of S at s to be the vector space m/m^2 (over K). Given $f \in \mathcal{a}$ we define $df_s \in T^*(S)_s$ to be the coset of $(f-f(s))$ modulo m^2 in m. We note that $f \mapsto df_s$ is a linear map of $\mathcal{a}$ onto $T^*(S)_s$ (since $f \mapsto f-f(s)$ is a linear map of $\mathcal{a}$ onto m).

1.9.5. Proposition. If $f, g \in \mathcal{a}$ then $d(fg)_s = f(s)dg_s+g(s)df_s$.

Proof. $fg-f(s)g(s) = f(s)(g-g(s))+g(s)(f-f(s))+(f-f(s))(g-g(s))$. Since $f-f(s)$ and $g-g(s)$ are both in m their product is in m^2. ■

1.9.6. Lemma. Let $f, g \in \mathcal{a}$ and suppose $df_s = dg_s$. Then $Df = Dg$ for all $D \in T(S)_s$.

Proof. $df_s = dg_s$ means that $f-f(s)$ differs from $g-g(s)$ by an element of m^2. Since D vanishes on elements of m^2 by 1.9.3,

$$D(f-g) = D((f-f(s))-(g-g(s))) = 0$$

so $Df = Dg$. ■

1.9.7. Theorem. Let $(S, \mathcal{a})$ be a ringed space over K and let $s \in S$. For each $D \in T(S)_x$ there is a unique linear functional ℓ_D on $T^*(S)_s$ such that $\ell_D(df_s) = Df$ for all $f \in \mathcal{A}$. Moreover, $D \mapsto \ell_D$ is a linear isomorphism of $T(S)_s$ with $T^*(S)_s^*$.

Proof. Since $f \mapsto df_s$ is a linear surjection of $\mathcal{A}$ onto $T^*(S)_s$ it follows from 1.9.6 that ℓ_D is a well-defined element of $T^*(S)_s^*$. Clearly $D \mapsto \ell_D$ is linear, and $\ell_D = 0$ obviously implies that $D = 0$, so it is injective. To see that $D \mapsto \ell_D$ is surjective, let $\ell \in T^*(S)_s^*$ and define a linear map $D : \mathcal{A} \to K$ by $Df = \ell(df_s)$. It is immediate from 1.9.5 that $D \in T(S)_s$, and clearly $\ell = \ell_D$.

∎

1.9.8. Corollary. There is a canonical linear injection $T^*(S)_s \to T(S)_s^*$; namely, if $\omega \in T^*(S)_s$, say, $\omega = df_s$, then we can consider ω as a linear functional on $T(S)_s$ by $\omega(D) = Df$. If $T(S)_s$ (or $T^*(S)_s$) is finite dimensional, this map is even a bijection.

Proof. Compose the reflexivity embedding $T^*(S)_s \to T^*(S)_s^{**}$, with the adjoint of the above isomorphism $T(S)_s \to T^*(S)_s^*$. ∎

1.9.9. Remark. Henceforth for $f \in \mathcal{A}$ and $s \in S$ we regard df_s as an element of $T(S)_s^*$.

1.9.10. Proposition. Let $f_1, \ldots, f_n \in \mathcal{A}$ and let f belong to the subalgebra $K[f_1, \ldots, f_n]$ of $\mathcal{A}$ generated by $f_1, \ldots, f_n$. Then for any $s \in S$, df_s is a linear combination of $d(f_1)_s, \ldots, d(f_n)_s$.

Proof. Let $h, g \in \mathcal{A}$ and suppose dh_s at dg_s are both linear combinations of $d(f_1)_s, \ldots, d(f_n)_s$. Then by 1.9.5 the same is true of $d(hg)_s$. It will clearly suffice to prove the proposition when f is a monomial $f_1^{j_1} f_2^{j_2} \dots f_n^{j_n}$. But using the above remark this follows by an obvious induction on the total degree $j_1 + \dots j_n$. ■

1.9.11. <u>Corollary</u>. If $\mathcal{A}$ is generated as a K algebra by a finite number (say n) of elements, then for each $s \in S$, $T(S)_s$ and $T^*(S)_s$ are finite dimensional (in fact of dimension $\leq n$). In particular (cf. 1.9.8) $T^*(S)_s$ is canonically isomorphic to $T(S)_s^*$.

1.9.12. <u>Remark.</u> Given a morphism of ringed spaces $\varphi : (S, \mathcal{A}) \to (U, B)$ and $s \in S$ let $M = \{F \in \mathcal{A} \mid F(s) = 0\}$ and $m = \{f \in B \mid f(\varphi(s)) = 0\}$. Clearly $\varphi^* : B \to \mathcal{A}$ $(f \mapsto f \circ \varphi)$ maps m into M and so induces a linear map of $m/m^2 = T^*U_{\varphi(s)}$ with $M/M^2 = T^*S_s$ which we denote by $T^*(\varphi)_s :$ $T^*U_{\varphi(s)} \to T^*S_s$. If $f \in B$ then it is immediate from the definition of $d(f \circ \varphi)_s$ that $d(f \circ \varphi)_s = T^*(\varphi)_s(df_{\varphi(s)})$. There is of course an adjoint linear map $T^*(\varphi)_s^* : (T^*S_s)^* \to (T^*U_{\varphi(s)})^*$. We claim that with the canonical isomorphisms $D \mapsto \ell_D$ of TS_s with $(T^*S_s)^*$ and $TU_{\varphi(s)}$ with $(T^*U_{\varphi(s)})^*$ (cf. 1.9.7) $T^*(\varphi)_s^*$ is identified with $T\varphi_s : TS_s \to TU_{\varphi(s)}$. For if $D \in TS_s$ and $f \in B$ then $T^*(\varphi)_s^*(\ell_D)(df_{\varphi(s)}) = \ell_D(T^*(\varphi)_s(df_{\varphi(s)})) = \ell_D(d(f \circ \varphi)_s) = D(f \circ \varphi) =$ $(T\varphi)_s(D)(f) = \ell_{\tilde{D}}(df_{\varphi(s)})$ where $\tilde{D} = (T\varphi)_s(D)$, and hence $T^*(\varphi)_s^*(\ell_D) = \ell_{\tilde{D}}$. Thus we have a commutative diagram:

$$\begin{array}{ccc} (T^*S_s)^* & \xrightarrow{T^*(\varphi)_s^*} & (T^*U_{\varphi(s)})^* \\ \ell \downarrow & & \downarrow \ell \\ TS_s & \xrightarrow[T\varphi_s]{} & TU_{\varphi(s)} \end{array}$$

as was to be shown.

1.9.13. Proposition. Let $(S, \mathcal{A})$ be a ringed space over K, X a Z-closed subspace of X and $\mathcal{I}(X) = \{f \in \mathcal{A} \mid (f|X) = 0\}$. Then TX_p is the annihilator of $\{df_p | f \in \mathcal{I}(X)\}$, i.e., TX_p = the set of $D \in TS_p$ such that $Df = df_p(D) = 0$ for all $f \in \mathcal{I}(X)$.

Proof. Recall that we identify TX_p with a subspace of TS_p via $T\varphi_p : TX_p \to TS_p$, where $\varphi : X \to S$ is inclusion, so $\varphi^* : \mathcal{A} \to \mathcal{A}(X)$ is restriction (cf. 1.9.2). That is if $f \in \mathcal{A}$ and $D \in TX_p$ then $df_p(D) = Df =$ $T\varphi_s(D)f = D(f \circ \varphi) = D(f|X)$. So if $f \in \mathcal{I}(X)$, i.e., $f|X = 0$, then $df_p(D) = 0$. Conversely if $D \in TY_p$ vanishes on $\{df_p | f \in \mathcal{I}(X)\}$, i.e., if $Df = 0$ for all $f \in \mathcal{I}(X)$, then $D : \mathcal{A} \to K$, being linear, is well defined mod $\mathcal{I}(X)$ and so is a linear map $D : \mathcal{A}/\mathcal{I}(X) \to K$. Identifying $\mathcal{A}(X)$ with $\mathcal{A}/\mathcal{I}(X)$ this is clearly an element of TX_p, giving D as above. ■

1.9.14. Corollary. If $\mathcal{I}(X)$ is generated (as an ideal) by $f_1, \ldots, f_k$, then

$$TX_p = \{D \in TS_p \mid (df_1)_p(D) = \ldots = (df_k)_p(D) = 0\}.$$

In particular, if $\mathcal{I}(X)$ is the principal ideal (f) then $TX_p = \{D \in TS_p | df_p(D) = 0\}$.

Proof. If $f \in \mathcal{I}(X)$ then $f = g_1 f_1 + \ldots + g_k f_k$. Then $df_p =$ $\sum_{i=1}^{k} (g_i(p)(df_i)_p + f_i(p)(dg_i)_p)$ by 1.9.5 and the linearity of d. Since $f_i \in \mathcal{I}(X)$ and $p \in X$, $f_i(p) = 0$ so df_p is a linear combination of the $(df_i)_p$. Then 1.9.14 is immediate from 1.9.13. ■

1.9.15. Definition. If $(S, \mathcal{Q})$ is a ringed space over K then a vector field on S is a derivation of the algebra $\mathcal{Q}$, i.e., a linear map $\partial : \mathcal{Q} \to \mathcal{Q}$ such that $\partial(fg) = (\partial f)g + f(\partial g)$. For any such ∂ and each $s \in S$ we define $\partial^s \in TS_s$, called the value of ∂ at s, by $\partial^s f = (\partial f)(s)$.

1.9.16. Remark. The set of derivations of any algebra is always a vector space. It is moveover a Lie algebra with the Lie bracket operation being the commutator. Thus we will speak of the Lie algebra of vector fields on a ringed space X.

1.9.17. Definition. Let X be a ringed space over X and let $t \mapsto \varphi_t$ be a one parameter group of automorphisms of X (cf. 1.5.28). We define a map $\partial : \mathcal{Q}(X) \to \mathcal{Q}(X)$ called the infinitesimal generator of $(t \mapsto \varphi_t)$ as follows: Let $\Phi : \mathcal{Q}(X) \mapsto \mathcal{Q}(X)[T]$ be the homomorphism such that $f \circ \varphi_t = \Phi(f)(t)$ for all $f \in \mathcal{Q}(X)$ at $t \in T$. Then for $f \in \mathcal{Q}(X)$, ∂f is the coefficient of T in $\Phi(f)$. Noting that $\Phi(f)(0) = f \circ \varphi_0 = f$ it follows that f is the constant term of $\Phi(f)$ so that $\Phi(f)-f$ is divisible by T and

$$\partial f = [(\Phi(f)-f)/T](0).$$

1.9.18. Proposition. Let $t \mapsto \varphi_t$ be a one parameter group of automorphisms of the ringed space X over K and let $\partial : \mathcal{Q}(X) \to \mathcal{Q}(X)$ be its infinitesimal generator. Then ∂ is a vector field on X.

Proof. Since $\Phi : \mathcal{Q}(X) \to \mathcal{Q}(X)[T]$ is a homomorphism of K algebras it is clear that $f \mapsto$ coefficient of T in $\Phi(f) = \partial f$ is linear. Given $f, g \in \mathcal{Q}(X)$, $\Phi(fg) = \Phi(f)\Phi(g)$ and hence $\Phi(fg) - fg = (\Phi(f)-f)\Phi(g) + (\Phi(g)-g)f$, so

$(\Phi(fg)-fg)/T = [(\Phi(f)-f)/T]\Phi(g)+[(\Phi(g)-g)/T]f$. Evaluating both sides at $T = 0$ (and recalling that $\Phi(g)(0) = g$) we get by the second formula for ∂ in 1.9.17:

$$\partial(fg) = (\partial f)g+(\partial g)f.$$ ∎

1.9.19. Definition. Let $\varphi : t \mapsto \varphi_t$ be a one parameter group of automorphisms of the ringed space X and let ∂ be its infinitesimal generator. An element f of $\mathcal{Q}(X)$ is called an invariant of φ if $f \circ \varphi_t = f$ for all $t \in K$ and f is called an infinitesimal invariant of φ if $\partial f = 0$.

1.9.20. Proposition. If φ is a one parameter group of automorphisms of the ringed space X then each invariant of φ is an infinitesimal invariant of φ.

Proof. If f is an invariant of φ then $\Phi(f)(t) = f \circ \varphi_t = f$ for all $t \in K$ and hence (cf. 1.5.28) $\Phi(f) = f$ so $\partial f = 0$. ∎

1.9.21. Proposition. Let S be a ringed space, $s \in S$, and U a Z-neighborhood of s in S. Let $M = \{F \in \mathcal{Q}(S) \mid F(s) = 0\}$ and $m = \{f \in \mathcal{Q}(U) \mid f(s) = 0\}$. If $F \in \mathcal{Q}(S)$ and $f = F|U$ then

(1) $F \in M$ if and only if $f \in m$.

(2) $F \in M^2$ if and only if $f \in m^2$.

Proof. Since $f(s) = F(s)$, (1) is trivial. If $F \in M^2$ then $F = \sum_{i=1}^{k} H_i K_i$ where $H_i, K_i \in M$. Putting $h_i = H_i|U$ and $k_i = K_i|U$, we have $h_i, k_i \in m$ by (1) and $f = \sum_{i=1}^{k} h_i k_i$ so $f \in m^2$. (Note that so far we have not used that U is a neighborhood of s.) Now suppose $f \in m^2$, say $f = \sum_{i=1}^{k} h_i k_i$ where $h_i, k_i \in m$. By definition of $\mathcal{Q}(U)$ (1.5.3) we have $h_i = H_i|U$ and $k_i = K_i|U$ where $H_i, K_i \in \mathcal{Q}(S)$ and by (1) $H_i, K_i \in M$. Note that $(F-\sum_i H_iK_i)|U = f-\sum_i h_ik_i = 0$. By 4) of 1.5.5, since U is a Z-neighborhood of s there exists G in $\mathcal{Q}(S)$ with $G(s) \neq 0$ and $G|(S-U) = 0$. Dividing by G(s) we can suppose $G(s) = 1$. Since G vanishes on S-U while $F-\sum_i H_iK_i$ vanishes on U we have $G(F-\sum_i H_iK_i) = 0$ and so

$$F = \sum_i H_iK_i+(1-G)(F-\sum_i H_iK_i).$$

Now F, (1-G), H_i, K_i all belong to M and it follows that $F \in M^2$. ■

1.9.22. Corollary. If $F_1, F_2 \in \mathcal{Q}(S)$ and $F_1|U = F_2|U$ then $(dF_1)_s = (dF_2)_s$ and hence (cf. 1.9.6) $DF_1 = DF_2$ for all $D \in TS_s$.

Proof. Since $(F_1-F_2)|U = 0$ we have by 1.9.21 that $(F_1-F_2) \in M^2$ and so $F_1-F_1(s)$ is equivalent to $F_2-F_2(s)$ modulo M^2. ■

1.9.23. Corollary. Let $\varphi : U \to S$ be the inclusion morphism, so that $\varphi^* : \mathcal{Q}(S) \to \mathcal{Q}(U)$ is the restriction homomorphism $F \mapsto F|U$. Then $T^*\varphi_s : T^*S_s \to T^*U_s$ is an isomorphism given explicitly by $dF_s \mapsto d(F|U)_s$. Also $T\varphi_s : TU_s \to TS_s$ is an isomorphism, given explicitly by $D \mapsto \tilde{D}$ where $\tilde{D}F = D(F|U)$.

Proof. Since $\varphi^* : \mathcal{Q}(S) \to \mathcal{Q}(U)$ is surjective (by definition of $\mathcal{Q}(U)$) and T^*U_s is the image of $\mathcal{Q}(U)$ under $f \mapsto df_s$ (1.9.4) it follows that $T^*\varphi_s$ is surjective (even when U is not necessarily a neighborhood of s), and by 1.9.21 it is immediate (again from 1.9.4) that $T^*\varphi_s$ is also injective when

U is a Z-neighborhood of s. Then 1.9.12 completes the proof. ■

1.9.24. Proposition. Let S be a ringed space and let $\tilde{S}$ be the regularization of S (cf. 1.5.16). If $s \in S$ then every element D of TS_s extends uniquely to an element $\tilde{D}$ of $T\tilde{S}_s$. If $h \in \mathcal{Q}(\tilde{S})$ and $h = f/g$ where $f, g \in \mathcal{Q}(S)$ then:

$$\tilde{D}h = (g(s)(Df)-f(s)(Dg))/g(s)^2.$$

Proof. We recall that $\tilde{S}$ has the same underlying set as S and that $\mathcal{Q}(\tilde{S}) = \mathcal{Q}_{reg}(S)$ is the set of quotients f/g where $f, g \in \mathcal{Q}(S)$ and $g(x) \neq 0$ for all $x \in S$. In particular $\mathcal{Q}(S)$ is a subalgebra of $\mathcal{Q}(\tilde{S})$.

If $\tilde{D}$ extends D then since $f = (f/g)g = hg$, $Df = \tilde{D}f = \tilde{D}(hg) = (\tilde{D}h)g(s)+(\tilde{D}g)h(s)$. Since $Dg = \tilde{D}g$ and $h(s) = f(s)/g(s)$ we get

$$(*) \qquad \tilde{D}h = (g(s)(Df)-f(s)(Dg))/g(s)^2.$$

Next suppose $h \in \mathcal{Q}(\tilde{S})$ and $h = f_1/g_1 = f_2/g_2$ where the f_i and g_i are in $\mathcal{Q}(S)$ and the g_i vanish nowhere on S. Then $f_1g_2 = f_2g_1$ so that $D(f_1g_2) = D(f_2g_1)$ or

$$(Df_1)g_2(s)+f_1(s)(Dg_2) = (Df_2)g_1(s)+f_2(s)(Dg_1)$$

and after some straightforward algebraic manipulation we get

$$(g_1(s)(Df_1)-f_1(s)(Dg_1))/g_1(s)^2 = (g_2(s)(Df_2)-f_2(s)(Dg_2))/g_2(s)^2$$

which proves that (*) gives a well-defined map $\tilde{D} : \mathcal{Q}(\tilde{S}) \to K$ extending $D : \mathcal{Q}(S) \to K$. We leave it to the reader to check that $\tilde{D}$ so defined has the formal properties of an element of $T\tilde{S}_s$. ■

1.9.25. Proposition. Let X be a ringed space, S a Z-open subset of X and $\widetilde{S}$ the regularization of S. The inclusion map $\varphi : \widetilde{S} \to X$ is a ringed space morphism and for each $s \in S$ $(T\varphi)_s : T\widetilde{S}_s \to TX_s$ is a linear isomorphism. The inverse isomorphism $D \mapsto \widetilde{D}$ of TX_s with $T\widetilde{S}_s$ is characterized as follows: if $h \in \mathcal{A}(\widetilde{S}) = \mathcal{A}_{reg}(S)$, say $h = f/g$ where $f, g \in \mathcal{A}(X)$ and g vanishes nowhere on S, then

$$\widetilde{D}h = (g(s)(Df) - f(s)(Dg))/g(s)^2.$$

Proof. Immediate from 1.9.23 and 1.9.24. ∎

1.9.26. Proposition. Let S_1 and S_2 be two Z-open neighborhoods of a point x_0 in a ringed space X and let $\widetilde{S}_i$ be the regularization of S_i and $\varphi_i : \widetilde{S}_i \to X$ the inclusion morphism. Let $D \in TX_{x_0}$ and let $D_i \in T(\widetilde{S}_i)_{x_0}$ be the unique solution of $(T\varphi_i)_{x_0}(D_i) = D$. Suppose $f_i \in \mathcal{A}(\widetilde{S}_i) = \mathcal{A}_{reg}(S_i)$ are such as to agree in some neighborhood U of s included in $S_1 \cap S_2$. Then $D_1 f_1 = D_2 f_2$.

Proof. Let $\widetilde{U}$ be the regularization of U, $\varphi : \widetilde{U} \to X$ the inclusion map and $\widetilde{D} \in T(\widetilde{U})_{x_0}$ the unique solution of $(T\varphi_i)_{x_0}(\widetilde{D}) = D$. Let $\psi_i : \widetilde{U} \to \widetilde{S}_i$ be inclusion morphisms. (Note: $\widetilde{U}$ is not a ringed subspace of $\widetilde{S}_i$; however, the restriction to U of a function regular on S_i is certainly regular on U, so ψ_i is a morphism.) Now clearly $\varphi = \varphi_i \circ \psi_i$ $(i = 1, 2)$ so that $D = (T\varphi)_{x_0}(\widetilde{D}) = (T\varphi_i)_{x_0}(T\psi_i)_{x_0}(\widetilde{D})$, and hence by the definition of D_i it follows that $D_i = (T\psi_i)_{x_0}(\widetilde{D})$. But then $D_i f_i = \widetilde{D}(f_i \circ \psi_i) = \widetilde{D}(f_i | U) = \widetilde{D}f$

where $f_1 | U = f_2 | U = f$. Hence $D_1 f_1 = D_2 f_2$. ∎

1.9.27. Proposition. If X and Y are ringed spaces, $x_0 \in X$ and $y_0 \in Y$ then $T(X \times Y)_{(x_0, y_0)}$ is canonically isomorphic to $TX_{x_0} \oplus TY_{y_0}$.

Proof. Given vector spaces V, V_1, V_2 and linear maps $\pi_i : V \to V_i$ and $j_i : V_i \to V$ such that $\pi_i \circ j_i$ are identity maps while $\pi_1 \circ j_2$ and $\pi_2 \circ j_1$ are zero maps we have a canonical identification of V with $V_1 \oplus V_2$. Let $j_{x_0} : Y \to X \times Y$ and $j_{y_0} : X \to X \times Y$ be the ringed space morphisms $y \mapsto (x_0, y)$ and $x \mapsto (x, y_0)$ and let $\pi_X : X \times Y \to X$ and $\pi_Y : X \times Y \to Y$ be the canonical projections. Then $\pi_x \circ j_{y_0} = \text{id}$ so $T(\pi_x)_{(x_0, y_0)} \circ T(j_{y_0})_{x_0}$ is the identity map of TX_{x_0} and similarly $T(\pi_Y)_{(x_0, y_0)} \circ T(j_{x_0})_{y_0}$ is the identity map of TY_{y_0}. Now $\pi_x \circ j_{x_0}$ is the constant map $y \mapsto x_0$ of Y into X. If $f \in \mathcal{Q}(X)$ then $f \circ (\pi_x \circ j_{x_0})$ is the constant function $y \mapsto f(x_0)$ on Y so for $D \in TY_{y_0}$, $D(f_0(\pi_x \circ j_{x_0})) = 0$; hence $(T\pi_x)_{(x_0, y_0)} \circ (Tj_{x_0})_{y_0} = 0$ and similarly $(T\pi_Y)_{(x_0, y_0)} \circ (Tj_{y_0})_{x_0} = 0$. ∎

1.9.28. Example. Let K be an infinite field and let $S = \prod_{j \in J} K$, so $\mathcal{Q}(S) = \mathcal{P}(\prod_{j \in J} K) = K[\{X_j\}_{j \in J}]$. We shall find the tangent space to S at an arbitrary point $s = \{s(j)\}_{j \in J}$. We consider first the case $s = 0$ (i.e., $s(j) = 0$ for all $j \in J$). Given $P = P(X)$ in $\mathcal{Q}(S)$, say

$$P = \sum_\alpha a_\alpha X^\alpha = a + \sum_{j \in J} a_{(j)} X_j + \sum_{|\alpha| \geq \alpha} a_\alpha X^\alpha$$

(where $|\alpha| = \sum_{j \in J} \alpha(j)$) clearly $P(0) = a$. Thus the ideal m of P vanishing at 0 is the ideal $(\{X_j\}_{j \in J})$ generated by the variables X_j, and so m^2 is the ideal $(\{X_i X_j\}_{(i,j) \in J \times J})$ generated by all monomials of degree 2. Since $(dP)_0$ is the coset of P-a in m modulo m^2 it follows that $(dP)_0 = \sum_{j \in J} a_{(j)}(dX_j)_0$ so that the $\{(dX_j)_0\}_{j \in J}$ span $m/m^2 = T^*S_0$. Moreover, they are linearly independent, for if $\sum_{j \in J} a_{(j)}(dX_j)_0 = 0$ then $\sum_{j \in J} a_{(j)} X_j \in m^2$ so we have a relation $\sum_{j \in J} a_{(j)} X_j = \sum_{|\alpha| \geq 2} a_\alpha X^\alpha$, and since a polynomial uniquely determines its coefficients the $a_{(j)}$ (and the a_α) are all zero. Thus $\{(dX_j)_0\}_{j \in J}$ is a basis for T^*S_0. It follows from 1. 9. 7 that we have an isomorphism $D \mapsto v_0$ of TS_0 with $\prod_{j \in J} K$, given explicitly by $v_D(j) = DX_j = (dX_j)_0(D)$. The inverse isomorphism $v \mapsto D^v$ of $\prod_{j \in J} K$ with TS_0 is given by

$$D^v(a + \sum_{j \in J} a_{(j)} X_j + \sum_{|\alpha| \geq 2} a_\alpha X^\alpha) = \sum_{j \in J} a_{(j)} v(j).$$

To compute TS_{v_0} at an arbitrary point $v_0 \in \prod_{j \in J} K$ we note that the map $v \mapsto v + v_0$ is a ringed space morphism $\tau_{v_0} : S \mapsto S$, i. e., if $P(X)$ is an element of $K[\{X_j\}_{j \in J}]$ then so is $P(X+v_0)$. In fact τ_{v_0} is an automorphism of S since its inverse is clearly τ_{-v_0}. It follows that $(T\tau_{v_0})_0$ maps TS_0 isomorphically onto TS_{v_0}. Thus $v \mapsto (T\tau_{v_0})_0(D^v)$ gives a canonical identification of $\prod_{j \in J} K$ with TS_{v_0}. We leave it to the reader to check that the value of $(T\tau_{v_0})_0(D^v)$ on a polynomial $P(X)$ is $\sum_{j \in J} (\partial P/\partial X_j)(v_0)v(j)$ where $(\partial P/\partial X_j)(X)$ is the formal partial derivative of $P(X)$ with respect to X_j, i. e.,

$\frac{\partial}{\partial x_j}(\sum_\alpha a_\alpha X^\alpha) = \sum_\alpha \alpha(j) a_\alpha X^{\alpha - 1_j}$ where $1_j(j') = 1$ if $j = j'$ and $1_j(j') = 0$ if $j \neq j'$. This can also be seen from the following remark. Given $v \in \prod_{j\in J} K$ consider the ring homomorphism

$$\Phi^v : \mathcal{P}(\prod_{j\in J} K) \to \mathcal{P}(\prod_{j\in J} K)[T]$$

defined by $P(X) \mapsto P(X+vT)$, i.e., $\sum_\alpha a_\alpha X^\alpha \mapsto \sum_\alpha a_\alpha (X+vT)^\alpha$ where $(X+vT)^\alpha = \prod_{j\in J}(X_j+v(j)T)^{\alpha(j)}$. Note that if we define $\varphi_t^v : X \to X$ by $\varphi_t^v = \tau_{tv}$ (where $t \in K$) then $P \circ \varphi_t^v = \Phi^v(P)(t)$, i.e., $(t, v_0) \mapsto \tau_{tv}(v_0) = v_0+tv$ is a one parameter group of automorphisms of $\prod_{j\in J} K$ (cf. 1.5.28). Let ∂^v denote the infinitesimal generator of this one parameter group (cf. 1.9.17). Thus $\partial^v : \mathcal{P}(\prod_{j\in J} K) \to \mathcal{P}(\prod_{j\in J} K)$ is a derivation defined by

$$\begin{aligned}\partial^v P(X) &= [(P(X+vT)-P(X))/T]_{T=0} \\ &= \sum_{j\in J} v(j)\frac{\partial P}{\partial x_j}(X).\end{aligned}$$

For each $v_0 \in S = \prod_{j\in J} K$ we get $\partial^v_{v_0} \in TS_{v_0}$ defined by $(\partial^v_{v_0} P) = (\partial^v P)(v_0) = \sum_j v(j)\frac{\partial P}{\partial x_j}(v_0)$. Thus the canonical isomorphism $\prod_{j\in J} K \to TS_{v_0}$ can also be described as the map $v \mapsto \partial^v_{v_0}$.

1.9.29. <u>Remark.</u> Let M be a C^k manifold $(k = 1, 2, \ldots, \infty, \omega)$. Then (cf. 1.6.4) we can regard M as a ringed space over $\mathbb{R}$ with structure ring $C^k(M) = C^k(M, \mathbb{R})$ and as such at each $p \in M$ we have a well-defined "algebraic" tangent space TM_p. On the other hand M also has a tangent space at p qua differentiable (i.e., C^1) manifold. We have already referred to and worked with this latter classical tangent space in Section 1.6 where it was also denoted by TM_p (this was before the algebraic tangent space was even defined). However, since our next goal is to relate these two distinct notions of tangent space we shall refer to the latter one as the "geometric" tangent space of M at p and denote it by T_gM_p. What we shall see is the following:

1) There is a natural linear map $\theta_p : T_gM_p \to TM_p$ (roughly speaking, for $f \in C^k(M)$, $\theta_p(v)f$ is the directional derivative of f at p in the direction v).

2) θ_p is always injective, so by using it as an identification we can regard T_gM_p as a linear subspace of TM_p (except for $k = \omega$ this is trivial).

3) For $k = \infty$ or $k = \omega$, θ_p is even bijective so we can in fact identify T_gM_p with TM_p. However, for $k < \infty$ whereas T_gM_p has the same dimension as M, TM_p has uncountable dimension (except in the trivial case $\dim M = 0$). Thus TM_p is not a useful object in these cases.

There is general disagreement on the "correct" way to define T_gM_p (e.g., 1-jets of smooth curves through p; the classical definition of a representative associated with each coordinate system and transforming in the appropriate way, etc.). Fortunately, all that matters are the functorial properties which characterize the association $(M, p) \mapsto T_gM_p$ up to canonical isomorphism. They can be stated as follows:

a) If φ is a C^1 map of a neighborhood $\mathcal{O}$ of p in M into a C^1 manifold N then there is associated a linear map $(T_g\varphi)_p : T_gM_p \to T_gN_p$. If $N = M$ and φ is an inclusion map then $(T_g\varphi)_p$ is the identity map.

b) $T_g(\varphi \circ \psi) = (T_g\varphi)_{\psi(p)} \circ (T_g\psi)_p$ (Chain rule).

c) If V is a finite dimensional vector space over $\mathbb{R}$ then for each $v \in V$ there is a canonical identification $j_v^V : T_g V_p \approx V$ such that if $\mathcal{O}$ is open in V and φ is a C^1 map of $\mathcal{O}$ onto another finite dimensional real vector space W over $\mathbb{R}$ then the following diagram commutes:

$$\begin{array}{ccc} T_g V_v & \xrightarrow{(T\varphi)_p} & T_g W_{\varphi(v)} \\ \downarrow j_v^V & & \downarrow j_{\varphi(v)}^W \\ V & \xrightarrow[(D\varphi)_p]{} & W \end{array}$$

where $(D\varphi)_p$ is the usual differential of φ at p.

The natural maps $\theta_p : T_g M_p \to TM_p$ can now be defined as follows: for the case M = V a finite dimensional real vector space over $\mathbb{R}$, $\theta_p(v)$ is the directional derivative at p in the direction $\tilde{v} = j_p^V(v)$, i.e., for $f \in C^k(V)$

$$\theta_p(v)f = \lim_{t\to 0} (1/t)(f(p+t\tilde{v})-f(p)).$$

In particular, for $V = \mathbb{R}^n$ and $\tilde{v} = (v_1, \ldots, v_n)$

$$\theta_p(v) = \sum_{i=1}^{n} v_i \frac{\partial}{\partial x_i}\Big|_p$$

where $x_1, \ldots, x_n$ are the standard coordinates for $\mathbb{R}^n$.

The definition of θ_p in the general case is determined by "naturality", i.e., the requirement that for $\mathcal{O}$ open in M and a C^k map $\varphi : \mathcal{O} \to N$ we should have commutativity in the diagram

$$\begin{array}{ccc} T_g M_p & \xrightarrow{\theta_p} & TM_p \\ (T_g \varphi)_p \downarrow & & \downarrow (T\varphi)_\varphi . \\ T_g N_{\varphi(p)} & \xrightarrow[\theta_{\varphi(p)}]{} & TN_{\varphi(p)} \end{array}$$

It is easily seen from the functorial properties of $(T_g\varphi)_p$ and the corresponding ones for $(T\varphi)_p$ (in particular the chain rule) that this definition is consistent (the crucial point in all this is of course the classical chain rule for the classical differential $D\varphi$). If $\varphi : \mathcal{O} \to \mathbb{R}^n$ is a chart for M with $p \in \mathcal{O}$, $v \in T_gM_p$, and $f \in C^k(M)$ then explicitly $\theta_p(v)f$ is the directional derivative of $\tilde{f} = f \circ \varphi^{-1}$ at $\tilde{p} = \varphi(p)$ in the direction $\tilde{v} = (T_g\varphi)_p(v)$.

<u>Theorem.</u> For all C^k manifolds M and $p \in M$ the canonical linear map $\theta_p : T_gM_p \to TM_p$ is injective.

<u>Proof.</u> Consider first the case $M = \mathbb{R}^N$. If $j_p^{\mathbb{R}^N}(v) = (v_1, \ldots, v_N) \in \mathbb{R}^N$ for some $v \in T\mathbb{R}^N_p$, then as has already been noted,

$$\theta_p(v) = \sum_{i=1}^{N} v_i \frac{\partial}{\partial x_i}\Big|_p$$ where $x_1, \ldots, x_n$ is the standard coordinate system for $\mathbb{R}^N$. In particular since $x_j \in C^k(\mathbb{R}^N)$ and $\theta_p(v)x_j = \sum_{i=1}^{N} v_i\delta_{ij} = v$, if $\theta_p(v) = 0$, then $j_p^{\mathbb{R}^N}(v) = 0$ so $v = 0$ and we have injectivity in this case. Now in general we can assume that M is a Z-closed, ringed subspace of some $\mathbb{R}^N$ simply by embedding M as a closed, regularly embedded C^k submanifold of $\mathbb{R}^N$ (cf. 1.6.6-1.6.9) so we have a commutative diagram

$$\begin{array}{ccc} T_gM_p & \xrightarrow{\theta_p} & TM_p \\ {\scriptstyle T_gi_p}\downarrow & & \downarrow{\scriptstyle Ti_p} \\ T_g\mathbb{R}^N_p & \xrightarrow[\theta_p]{} & T\mathbb{R}^N_p \end{array} .$$

Now the bottom arrow has just been shown to be a monomorphism, and the left vertical arrow is injective by definition of an embedding. It follows that the top arrow must also be injective. ∎

Corollary. $\dim(TM_p) \geq \dim M$, with equality if and only if $\theta_p : T_gM_p \to TM_p$ is bijective.

Proof. Immediate from the theorem and the fact that $\dim(T_gM_p) = \dim M$. ∎

1.9.29. Example. Let $S = \mathbb{R}^n$, considered as a ringed space over $\mathbb{R}$ with structure ring $\mathcal{Q}(S) = C^k(\mathbb{R}^n)$ with k in the set $\{0,1,2,\ldots,\infty,\omega\}$. We shall next find the tangent space $T\mathbb{R}^n_x$ (or at least its dimension) at a point $x \in \mathbb{R}^n$. Since in every case $v \mapsto v+x$ is an automorphism $\tau_x : \mathbb{R}^n \approx \mathbb{R}^n$, $(T\tau_x)_0$ maps $T\mathbb{R}^n_0$ isomorphically onto $T\mathbb{R}^n_x$, so we can assume $x = 0$. We shall consider three cases separately, namely $k = 0$, k a positive integer, and $k = \infty$ or ω. The results are slightly surprising. In the first case $T\mathbb{R}^n_0 = (0)$, in the second case $T\mathbb{R}^n_0$ has uncountable dimension, and only in the third case do we get the "expected" result, namely $T\mathbb{R}^n_0 \approx \mathbb{R}^n$.

Case I. $(k = 0)$. Let m denote the ideal of functions in $C^0(\mathbb{R}^n)$ vanishing at 0. We shall show that $m = m^2$ so that $T^*\mathbb{R}^n_0 = m/m^2 = (0)$ and hence $T\mathbb{R}^n_0 = (T^*\mathbb{R}^n_0)^* = (0)$. Let $f \in m$. If f is everywhere non-negative then $g = \sqrt{f} \in C^0(\mathbb{R}^n)$ and clearly $g \in m$, so $f = g^2 \in m^2$. In general we can write $f = f^+ - f^-$ where f^+, f^- are everywhere non-negative elements of m (namely $f^+(x) = f(x)$ where $f(x) \geq 0$ and $f^+(x) = 0$ elsewhere and $f^-(x) = -f(x)$ where $f(x) \leq 0$ and $f^-(x) = 0$ elsewhere). Then f^+ and f^- and hence f are in m^2. Note that this same argument works if we replace $\mathbb{R}^n$ by an arbitrary topological space.

Case II. (k a positive integer). In this case we show that $T^*\mathbb{R}^n_0$ ($= m/m^2$) and hence its dual $T\mathbb{R}^n_0$, have uncountable dimension. We note that it suffices to consider the case $n = 1$. For consider the embedding $j : \mathbb{R} \to \mathbb{R}^n$, $x \mapsto (x, 0, \ldots, 0)$ and the projection $\pi : \mathbb{R}^n \to \mathbb{R}$, $(x_1, \ldots, x_n) \mapsto x_1$. These are clearly C^k maps, hence ringed space morphisms, and $\pi \circ j$ is the identity map of $\mathbb{R}$ so $(T\pi)_0 \circ (Tj)_0$ is the identity map of $T\mathbb{R}_0$. Hence Tj_0 maps $T\mathbb{R}_0$ isomorphically onto a linear subspace of $T\mathbb{R}^n_0$, so if $T\mathbb{R}_0$ has uncountable dimension so does $T\mathbb{R}^n_0$. We note that the set of functions $\{f_\lambda\}$, $k < \lambda < k+1$, defined by $f_\lambda(x) = x^\lambda$, all belong to m, so it will suffice to show that they are linearly independent modulo m^2. Suppose then that $f = \sum_{i=1}^m a_i f_{\lambda_i}$ (where the λ_i are distinct) belongs to m^2. Note that

$$f^{(k)}(x) = \sum_i k!\, a_i x^{\lambda_i - k}$$

so that $f^{(k)}(0) = 0$ and $(f^{(k)}(x) - f^{(k)}(0))/x = \sum_{i=1}^m k!\, a_i x^{\lambda_i - k - 1}$. Since $\lambda_i < k+1$ and the λ_i are distinct it follows that if $f^{(k)}$ has a derivative at 0 then all the a_i must be zero. Thus it will suffice to prove that whenever $f \in m^2$, $f^{(k)}$ has a derivative at 0. We can assume that $f = gh$ where $g, h \in m$ since in any case f is a finite sum of such products. Now a simple induction shows that for $0 \leq j \leq k$ we have $f^{(j)} = g^{(j)}h + gh^{(j)} + \ell_j$ where $\ell_j \in C^{k-j+1}$. (In fact $\ell_0 = 0$ and $\ell_{j+1} = \ell_j + g^{(j)}h' + g'h^{(j)}$). In particular $f^{(k)} = g^{(k)}h + gh^{(k)} + \ell_k$ where $\ell_k \in C^1$ Thus we are reduced to verifying that the product of a differentiable function which vanishes at zero with an arbitrary continuous function has a derivative at zero. But that is trivial.

Case III. ($k = \infty$ or $k = \omega$). In this case we show that $(dx_1)_0, \ldots, (dx_n)_0$ is a basis for $m/m^2 = T^*\mathbb{R}^n_0$. We use the fact that if $g \in C^k(\mathbb{R}^n)$ then $g_i = \partial g/\partial x_i \in C^k(\mathbb{R}^n)$, $i = 1, \ldots, n$, and $\bar{g} \in C^k(\mathbb{R}^n)$ where $\bar{g}(x_1, \ldots, x_n) = \int_0^1 g(tx_1, \ldots, tx_n)dt$. We note that for any $f \in C^k(\mathbb{R}^n)$

$$f(x)-f(0) = \int_0^1 \frac{d}{dt} f(tx)dt = \sum_{i=1}^n x_i \bar{f}_i(x),$$

so that if $f \in m$ (i.e., $f(0) = 0$) then f is in the ideal of $C^k(\mathbb{R}^n)$ generated by $x_1, \ldots, x_n$, and the converse is obvious. It follows of course that m^2 is generated by the monomials $x_i x_j$. It also follows that $T^*\mathbb{R}^n_0$ is spanned by $(dx_1)_0, \ldots, (dx_n)_0$. Suppose then $\sum_{i=1}^n a_i (dx_i)_0 = 0$, i.e., $\sum_{i=1}^n a_i x_i \in m^2$, say $\sum_{i=1}^n a_i x_i = \sum_{ij} x_i x_j g_{ij}(x)$ where $g_{ij} \in C^k(\mathbb{R}^n)$ and in particular g_{ij} is continuous at 0. Putting $x_i = t$ and $x_j = 0$ for $j \neq i$ we see that $ta_i = t^2 g_{ii}(0, \ldots, t, \ldots, 0)$. Dividing by t and letting t approach zero we see that $a_i = 0$ so that the $(dx_i)_0$ are linearly independent.

1.9.30. Theorem. If $k = \infty$ or $k = \omega$ then for any C^k manifold M and $p \in M$ the natural map (cf. 1.9.28) $\theta_p : T_g M_p \to TM_p$ of the geometric tangent space at p into the algebraic tangent space at p is bijective.

Proof. Recall that by the corollary of 1.9.28 we only have to show that TM_p has the same dimension as M. In particular, by 1.9.29 the theorem is true for $M = \mathbb{R}^n$ so for $\mathbb{R}^n$ we do not have to distinguish between $T_g\mathbb{R}^n_p$ and $T\mathbb{R}^n_p$ and similarly for their duals. Now by 1.6.6-1.6.9 we can assume M is embedded as a closed, regularly embedded C^k submanifold of $\mathbb{R}^n$ and is thereby a Z-closed, ringed subspace of $\mathbb{R}^n$. Let $I = \{f \in C^k(\mathbb{R}^n) \mid (f|M) = 0\}$

and let $V_p = \{df_p \mid f \in I\} \subseteq T\mathbb{R}^{n*}_p$ and $\rho = \dim V_p$. By 1.9.12 TM_p is the annihilator of V_p and hence $\dim(TM_p) = \dim(T\mathbb{R}^{n*}_p) - \rho = n - \rho$. But by 1.6.14 $\rho = \dim \mathbb{R}^n - \dim M$ so that $\dim(TM_p) = \dim M$. ■

1.9.31. Remark. Let M be a C^ω manifold and let $\widetilde{M}$ denote M considered as a C^∞ manifold. Since $C^\omega(M) \subseteq C^\infty(\widetilde{M})$ the identity map $j : \widetilde{M} \to M$ is a ringed space morphism. Since it is trivial that $(T_g j)_p : T_g\widetilde{M}_p \to T_g M_p$ is an isomorphism it follows from 1.9.30 that $(Tj)_p : T\widetilde{M}_p \to TM_p$ is also an isomorphism. In case $M = \mathbb{R}^n$ we can denote by $\hat{M}$ the ringed space $\mathbb{R}^n$ with structure ring the polynomial ring $\mathcal{P}(\mathbb{R}^n)$ and now we have ringed space morphisms $\hat{M} \to \widetilde{M} \to M$ where $i : \hat{M} \to \widetilde{M}$ is again the identity map. Since the standard coordinates $x_1, \ldots, x_n$ for $\mathbb{R}^n$ are included in $\mathcal{P}(\mathbb{R}^n)$ it follows from 1.9.27 that $(Ti)_p : T\hat{M}_p \to T\widetilde{M}_p$ is also bijective. Much later we shall see that this holds for any non-singular real algebraic affine variety M.

1.10. The Local Ring of a Point.

Let X be a topological space and $x_0 \in X$. Let $\mathcal{F}_{X,x_0}$ denote the set of all functions $f : U \to K$ where U is some neighborhood of x_0 (depending on f). Given $f_1 : U_1 \to K$ and $f_2 : U_2 \to K$ in $\mathcal{F}_{X,x_0}$ we say that f_1 and f_2 have the same germ at x_0 if there is a neighborhood V of x_0, $V \subseteq U_1 \cap U_2$, such that $f_1 | V = f_2 | V$. This is clearly an equivalence relation on $\mathcal{F}_{X,x_0}$ whose set of equivalence classes we denote by G_{X,x_0} and call the set of germs of K-valued functions at x_0 in X. The equivalence class of f is denoted by $[f]_{x_0}$ and called the germ of f at x_0. If $\alpha \in K$ we define $\alpha[f]_{x_0} = [\alpha f]_{x_0}$. Given f_1 and f_2 as above we define $[f_1]_{x_0} + [f_2]_{x_0} = [f_1 + f_2]_{x_0}$ and $[f_1]_{x_0}[f_2]_{x_0} = [f_1 f_2]_{x_0}$ where $f_1 + f_2$ and $f_1 f_2$ are the obvious maps of $U_1 \cap U_2$ into K. We leave it to the reader to check that these operations on G_{X,x_0} are well defined (i.e., depend only on the germs $[f]_{x_0}$ and not on the representatives f) and give G_{X,x_0} the structure of a K-algebra. Moreover, we have a well-defined homomorphism $[f]_{x_0} \mapsto f(x_0)$ of G_{X,x_0} onto K, called evaluation. If $\gamma = [f]_{x_0}$ we write $\gamma(x_0) = f(x_0)$.

Next suppose that for each open set $\mathcal{O}$ of X we have associated an algebra $\mathcal{R}(\mathcal{O})$ of K-valued functions on $\mathcal{O}$. Suppose moreover that if U is an open set included in $\mathcal{O}$ then the inclusion homomorphism $f \mapsto f|U$ maps $\mathcal{R}(\mathcal{O})$ into $\mathcal{R}(U)$. Let $G^{\mathcal{R}}_{X,x_0}$ denote the set of $[f]_{x_0}$ in G_{X,x_0} where $f \in \mathcal{R}(\mathcal{O})$ for some open set $\mathcal{O}$ containing x_0. It is easily seen that $G^{\mathcal{R}}_{X,x_0}$

is a subalgebra of G_{X,x_0}. In fact if we regard the collection of open neighborhoods of x_0 in X as a directed set under inclusion then $\{\mathcal{R}(\mathcal{O})\}$ together with the restriction homomorphisms is a direct system of K-algebras and $G^{\mathcal{R}}_{X,x_0}$ is just the direct limit $\varinjlim \mathcal{R}(\mathcal{O})$. If we now specialize one step further and assume X is a ringed space over K with its Z-topology and let $\mathcal{R}(\mathcal{O})$ be $\mathcal{Q}_{reg}(\mathcal{O})$ then $G^{\mathcal{R}}_{X,x_0}$ is what is called the local ring of X at x_0, denoted by $\mathcal{O}_{X,x_0}$.

1.10.1. <u>Definition.</u> Let X be a ringed space over K. For each $x_0 \in X$ we define the <u>local ring of</u> X <u>at</u> x_0, denoted by $\mathcal{O}_{X,x_0}$, to be the algebra of germs $[f]_{x_0}$ of K-valued functions at x_0, where $f : \mathcal{O} \to K$ is an element of $\mathcal{Q}_{reg}(\mathcal{O})$, $\mathcal{O}$ being some Z-open neighborhood of x_0 (depending on f), i.e., f is of the form $x \mapsto h(x)/g(x)$ where $h, g \in \mathcal{Q}(X)$ and $g(x) \neq 0$ for all $x \in \mathcal{O}$; see above for details. The maximal ideal of $\mathcal{O}_{X,x_0}$, consisting of γ such that $\gamma(x_0) = 0$, will be denoted by M_{X,x_0}.

1.10.12. <u>Remark.</u> We recall that a commutative ring R (with identity) is called a <u>local ring</u> if it has a unique maximal ideal M. If r is any non-unit of R then the principal ideal $(r) = Rr$ is proper. Since by Zorn's lemma any proper ideal of a ring with identity is included in a maximal ideal, it follows that when R is a local ring its maximal ideal M consists of exactly the non-units of R (a unit can of course never be an element of a proper ideal). Conversely if R is a commutative ring with identity in which the set of non-units is an ideal M, then (again because proper ideals consist of non-units) M is actually a maximum ideal of R, i.e., includes every proper ideal of R, so in particular it is the unique maximal ideal of R and R is a local ring.

We note that if X is a ringed space and $x_0 \in X$ then $\mathcal{O}_{X,x_0}$, the local ring of X at x_0, is a local ring in the above sense. For suppose $\gamma \in \mathcal{O}_{X,x_0}$, say $\gamma = [f]_{x_0}$ where $f \in \mathcal{Q}_{reg}(\mathcal{O})$, $\mathcal{O}$ being a Z-open neighborhood of x_0 in X. Then $f(x) = g(x)/h(x)$ where $g, h \in \mathcal{Q}(X)$ and $h(x) \neq 0$ for $x \in \mathcal{O}$. Suppose $\gamma \notin M_{X,x_0}$, i.e., $0 \neq \gamma(x_0) = g(x_0)/h(x_0)$ so $g(x_0) \neq 0$. Let $U = \{x \in \mathcal{O} \mid g(x) \neq 0\}$ so U is a Z-open neighborhood of x_0 in X and $x \mapsto h(x)/g(x)$ is an element of $\mathcal{Q}_{reg}(U)$. Then $[h/g]_{x_0} \in \mathcal{O}_{X,x_0}$ is clearly an inverse for γ in $\mathcal{O}_{X,x_0}$ so γ is a unit of $\mathcal{O}_{X,x_0}$ and all non-units of $\mathcal{O}_{X,x_0}$ are contained in M_{X,x_0}.

If R_1 and R_2 are two local rings with maximal ideals M_1 and M_2 respectively, then a ring homomorphism $\varphi : R_1 \to R_2$ is called a morphism of local rings if $\varphi(M_1) \subseteq M_2$. Note that if R is a local ring with maximal ideal M and $\mathcal{J}$ is any ideal of R (so $\mathcal{J} \subseteq M$) then $R/\mathcal{J}$ is a local ring with maximal ideal $M/\mathcal{J}$ (for the ideals of $R/\mathcal{J}$ are in one to one inclusion preserving correspondence with the ideals J of R including $\mathcal{J}$, under the map $J \mapsto J/\mathcal{J}$). In particular with the morphism of local rings $\varphi : R_1 \to R_2$ above we have associated the ideal $R_2\varphi(M_1) = \mathcal{J} \subseteq M_2$ and hence the local ring $R_2/(R_2\varphi(M_1))$.

1.10.3. <u>Proposition</u>. Let X be a ringed space over K and $x_0 \in X$. Given

$f_1, g_1 \in \mathcal{Q}(X)$ with $g_1(x_0) \neq 0$, $[f_1]_{x_0}/[g_1]_{x_0}$ is a well-defined element of $\mathcal{O}_{X,x_0}$; moreover every element of $\mathcal{O}_{X,x_0}$ can be represented in this form. If $f_2 g_2 \in \mathcal{Q}(X)$ and $g_2(x_0) \neq 0$ then $[f_1]_{x_0}/[g_1]_{x_0} = [f_2]_{x_0}/[g_2]_{x_0}$ if and only if there exists $h \in \mathcal{Q}(X)$ with $h(x_0) \neq 0$ and

$$h(f_1 g_2 - f_2 g_1) = 0.$$

Proof. Since $g_1(x_0) \neq 0$, $[g_1]_{x_0} \notin M_{X,x_0}$ so by 1.10.2 $[g_1]_{x_0}$ is a unit of $\mathcal{O}_{X,x_0}$ and $[f_1]_{x_0}/[g_1]_{x_0}$ is a well-defined element of $\mathcal{O}_{X,x_0}$. Let $\gamma \in \mathcal{O}_{X,x_0}$, say $\gamma = [f/g]_{x_0}$ where $f, g \in \mathcal{Q}(\mathcal{O})$, $\mathcal{O}$ is a Z-open neighborhood of x_0, and g does not vanish in $\mathcal{O}$. Since $g(f/g) = f$ we have $[g]_{x_0}[f/g]_{x_0} = [f]_{x_0}$ and hence $\gamma = [f/g]_{x_0} = [f]_{x_0}/[g]_{x_0}$. Choose $f_1, g_1 \in \mathcal{Q}(X)$ such that $f = f_1|\mathcal{O}$ and $g = g_1|\mathcal{O}$. Since $\mathcal{O}$ is a neighborhood of x_0, $[f] = [f_1]_{x_0}$ and $[g] = [g_1]_{x_0}$ so $\gamma = [f_1]_{x_0}/[g_1]_{x_0}$.

If $h(f_1 g_2 - f_2 g_1) = 0$ where $h(x_0) \neq 0$ then $U = \{x \in X | h(x) \neq 0\}$ is a Z-open neighborhood of x_0 in X and in U, $f_1 g_2 = f_2 g_1$ so $[f_1]_{x_0}[g_2]_{x_0} = [f_2]_{x_0}[g_1]_{x_0}$ so $[f_1]_{x_0}/[g_1]_{x_0} = [f_2]_{x_0}/[g_2]_{x_0}$. Conversely if the latter equality holds then $[f_1 g_2 - f_2 g_1]_{x_0} = 0$ which means $(f_1 g_2 - f_2 g_1)(x) = 0$ for all x in some neighborhood U of x_0. By 4) of 1.5.5 there is an $h \in \mathcal{Q}(X)$ such that $h(x_0) \neq 0$ and $\{x \in X | h(x) \neq 0\} \subseteq U$. Clearly $h(f_1 g_2 - f_2 g_1) = 0$. ■

1.10.4. Remark. The local ring $\mathcal{O}_{X,x_0}$ is functorial in the following precise sense. Let $\varphi: Y \to X$ be a ringed space morphism, $y_0 \in Y$ and $\varphi(y_0) = x_0$. Then we have a morphism of local rings

$$\mathcal{O}_{\varphi, y_0} : \mathcal{O}_{X, x_0} \to \mathcal{O}_{Y, y_0}$$

which we frequently will denote simply by φ^*. It is defined as follows: let $\gamma = [f]_{x_0} \in \mathcal{O}_{X, x_0}$, say $f \in \mathcal{Q}_{reg}(\mathcal{O})$ where $\mathcal{O}$ is a Z-open neighborhood of x_0 in X. Then $\varphi^{-1}(\mathcal{O})$ is a Z-open neighborhood of y_0 in Y and $f \circ \varphi^{-1} \in \mathcal{Q}_{reg}(\varphi^{-1}(\mathcal{O}))$; we define $\varphi^* \gamma = [f \circ \varphi^{-1}]_{y_0}$. Note that $(\varphi^* \gamma)(y_0) = \gamma(x_0)$ so in particular $\gamma(x_0) = 0$ implies $\varphi^*(\gamma)(y_0) = 0$, i.e., $\varphi^*(M_{X, x_0}) \subseteq M_{Y, y_0}$, as required for a morphism of local rings. If $\mathcal{J} = \mathcal{O}_{Y, y_0} \varphi^*(M_{X, x_0})$ then the quotient local ring $\mathcal{O}_{Y, y_0}/\mathcal{J}$ is called <u>the local ring of</u> φ <u>at</u> y_0.

1.10.5. <u>Remark.</u> Let X be a ringed space, U a Z-open set in X and $x_0 \in U$. Then we have a natural homomorphism $j_{U, x_0} : \mathcal{Q}_{reg}(U) \to \mathcal{O}_{X, x_0}$, namely, $f \mapsto [f]_{x_0}$. We will frequently just write j_{x_0} or even j instead of j_{X, x_0}. Here natural means the following. If $\varphi : Y \to X$ is a ringed space morphism, $y_0 \in Y$ and $\varphi(y_0) = x_0$, then $V = \varphi^{-1}(U)$ is a Z-open subset of Y containing y_0 and we have $j_{V, y_0} : \mathcal{Q}_{reg}(V) \to \mathcal{O}_{Y, y_0}$. Moreover we have a ring homomorphism $\varphi^* : \mathcal{Q}_{reg}(U) \to \mathcal{Q}_{reg}(V)$ given by $f \mapsto f \circ \varphi$, and a local ring homomorphism $\varphi^* = \mathcal{O}_{\varphi, y_0} : \mathcal{O}_{Y, y_0} \to \mathcal{O}_{X, x_0}$ (cf. 1.10.4). Clearly we have a commutative diagram:

$$\begin{array}{ccc}
\mathcal{A}_{reg}(V) & \xrightarrow{\varphi^*} & \mathcal{A}_{reg}(U) \\
{\scriptstyle j_{V,y_0}}\downarrow & & \downarrow{\scriptstyle j_{U,x_0}} \\
\mathcal{O}_{Y,y_0} & \xrightarrow[\varphi^*]{} & \mathcal{O}_{X,x_0}
\end{array}$$

1.10.6. Definition. If X is a ringed space over K and $x_0 \in X$ we define K vector spaces T_{X,x_0} and T^*_{X,x_0} as follows: T_{X,x_0} is the vector space of K-linear maps $D : \mathcal{O}_{X,x_0} \to K$ satisfying:

$$D(\lambda\mu) = (D\lambda)\mu(x_0)+\lambda(x_0)(D\mu);$$

$T^*_{X,x_0} = M_{X,x_0}/M^2_{X,x_0}$.

If $\gamma \in M_{X,x_0}$ we denote its equivalence class modulo M^2_{X,x_0} by $d\gamma \in T^*_{X,x_0}$. More generally if $\gamma \in \mathcal{O}_{X,x_0}$, so $\gamma-\gamma(x_0) \in M_{X,x_0}$, we denote the equivalence class of $\gamma-\gamma(x_0)$ by $d\gamma$ and we note that $\gamma \mapsto d\gamma$ is a surjective linear map $\mathcal{O}_{X,x_0} \to T^*_{X,x_0}$.

Both T_{X,x_0} and T^*_{X,x_0} are functorial (the former is covariant, the latter contravariant). Given a ringed space morphism $\varphi : Y \to X$ and $y_0 \in Y$ with $\varphi(y_0) = x_0$ we have $T_{\varphi,y_0} : T_{Y,y_0} \to T_{X,x_0}$ defined by $T_{\varphi,y_0}(D) = D \circ \mathcal{O}_{\varphi,y_0} : \mathcal{O}_{X,x_0} \to \mathcal{O}_{Y,y_0} \xrightarrow{D} K$, or explicitly $T_{\varphi,y_0}(D)[f]_{x_0} = D([f \circ \varphi]_{y_0})$. And we have $T^*_{\varphi,y_0} : T^*_{X,x_0} \to T^*_{Y,y_0}$, the obvious map of $M_{X,x_0}/M^2_{X,x_0} \to M_{Y,y_0}/M^2_{Y,y_0}$ induced by the local ring morphism $\mathcal{O}_{\varphi,y_0} = \varphi^* : \mathcal{O}_{X,x_0} \to \mathcal{O}_{Y,y_0}$ which maps $M_{X,x_0} \to M_{Y,y_0}$ and hence $M^2_{X,x_0} \to M^2_{Y,y_0}$. Note that by definition we have:

$$T^*_{\varphi, y_0}(d\gamma) = d(\varphi^* \gamma)$$

for $\gamma \in \mathcal{O}_{X,x_0}$, where $\varphi^* = \mathcal{O}_{\varphi, y_0}$.

1.10.7. Proposition. If X is a ringed space over K and $x_0 \in X$ then there is a canonical isomorphism of the dual of T^*_{X,x_0} with T_{X,x_0}. If $D \in T_{X,x_0}$ and $\gamma \in \mathcal{O}_{X,x_0}$ then the value of D (considered as a linear functional on T^*_{X,x_0}) at $d\gamma$, which we denote by $d\gamma(D)$, is just $D\gamma$ and this characterizes the isomorphism. The isomorphism is natural in the sense that the following diagram commutes (for a ringed space morphism $\varphi : Y \to X$ with $\varphi(y_0) = x_0$):

$$\begin{array}{ccc} (T^*_{X,x_0})^* & \xrightarrow{(T^*_{\varphi, y_0})^*} & (T^*_{Y,y_0})^* \\ \wr\wr & & \wr\wr \\ T_{X,x_0} & \xrightarrow[T_{X,x_0}]{} & T_{Y,y_0} \end{array} .$$

Proof. The same as 1.9.7 and 1.9.12, mutatis mutandis. ■

1.10.8. Remark. Henceforth we shall tacitly identify T_{X,x_0} with the dual of T^*_{X,x_0}, and dually we identify T^*_{X,x_0} with a subspace of $(T_{X,x_0})^*$.

1.10.9. Definition. Let X be a ringed space over K, $x_0 \in X$. Let m denote the ideal of function in $\mathcal{Q}(X)$ vanishing at x_0. The natural homomorphism $j_{x_0} : \mathcal{Q}(X) \to \mathcal{O}_{X,x_0}$ clearly maps m into M_{X,x_0} and therefore induces a linear map of $m/m^2 = T^*X_{x_0}$ into $M_{X,x_0}/M^2_{X,x_0} = T^*_{X,x_0}$. We call this the canonical map $T^*X_{x_0} \to T^*_{X,x_0}$ and note that it is given by $df_{x_0} \mapsto d[f]_{x_0}$ for $f \in \mathcal{Q}(X)$.

Dually we have a canonical map $T_{X,x_0} \to TX_{x_0}$ defined by $D \mapsto D \circ j_{x_0}$, i.e., $D \in T_{X,x_0}$ is mapped to $\overline{D} \in TX_{x_0}$ where for $f \in \mathcal{Q}(X)$

$$\overline{D}f = D([f]_{x_0}).$$

1.10.10. Remark. The reader should check that these canonical maps are indeed natural. That is, if $\varphi : Y \to X$ is a ringed space morphism with $\varphi(y_0) = x_0$ then we have induced maps $T^*_{\varphi,y_0} : T^*_{X,x_0} \to T^*_{Y,y_0}$ and $T^*(\varphi)_{y_0} : T^*X_{x_0} \to T^*Y_{y_0}$ and these commute with the canonical maps; similarly the induced maps $(T\varphi)_{y_0} : TY_{y_0} \to TX_{x_0}$ and $T_{\varphi,y_0} : T_{Y,y_0} \to T_{X,x_0}$ commute with the canonical maps. The reader should also note that the canonical maps commute, in the obvious sense, with the identifications of T_{X,x_0} with $(T^*_{X,x_0})^*$ and TX_{x_0} with $(T^*X_{x_0})^*$.

A priori it is not clear that the above canonical map are either injective or surjective. In fact they are isomorphisms, a very important fact whose proof we have been building up to for some time.

1.10.11. Theorem. Let X be a ringed space over K and $x_0 \in X$. The canonical linear map $T_{X,x_0} \to TX_{x_0}$ (cf. 1.10.9) is an isomorphism of vector spaces. The inverse isomorphism $D \mapsto \widetilde{D}$ of TX_{x_0} with T_{X,x_0} is characterized as follows: let $\gamma \in \mathcal{O}_{X,x_0}$ and express γ in the form $[f]_{x_0}/[g]_{x_0}$ for some $f, g \in \mathcal{A}(X)$ with $g(x_0) \neq 0$ (cf. 1.10.3), then

$$\widetilde{D}\gamma = (g(x_0)(Df)-f(x_0)(Dg))/g(x_0)^2.$$

<u>Proof.</u> Given $D \in TX_{x_0}$ we define $\widetilde{D} : \mathcal{O}_{X,x_0} \to K$ as follows: let $\gamma \in \mathcal{O}_{X,x_0}$, say $\gamma = [h]_{x_0}$ where $h \in \mathcal{A}_{reg}(S)$, S being some open neighborhood of x_0 in X; let $\widetilde{S}$ be the regularization of S and let $\overline{D} : \mathcal{A}_{reg}(S) \to K$ be the unique element of $T\widetilde{S}_{x_0}$ satisfying $(T\varphi)_{x_0}(\overline{D}) = D$, where $\varphi : \widetilde{S} \to X$ is the inclusion map (cf. 1.9.25). Then define $\widetilde{D}\gamma = \overline{D}h$. It is precisely the context of 1.9.26 that $\widetilde{D}$ depends only on γ and not on the representative h. If $\mu \in \mathcal{O}_{X,x_0}$ then we can get a representative $k : S' \to K$ for μ, and replacing S and S' by $S \cap S'$ (and calling the latter S) we can suppose h and k are both regular on the same set S. Since $\overline{D}(h+k) = \overline{D}h+\overline{D}k$ and $\overline{D}(hk) = (\overline{D}h)k(x_0)+h(x_0)(\overline{D}k)$ we get the corresponding results for $\widetilde{D}(\gamma+\mu)$ and $\widetilde{D}(\gamma\mu)$ which show $\widetilde{D} \in T_{X,x_0}$.

Suppose $\gamma = [f]_{x_0}/[g]_{x_0}$ where $f, g \in \mathcal{A}(X)$ and $g(x_0) \neq 0$. Then letting $S = \{x \in X \mid g(x) \neq 0\}$ and defining $h \in \mathcal{A}_{reg}(S)$ by $h(x) = f(x)/g(x)$ for $x \in S$, we have $[h]_{x_0} = [f]_{x_0}/[g]_{x_0} = \gamma$ so that $\widetilde{D}\gamma = \overline{D}h$ where $\overline{D}$ is as defined above. By 1.9.25 we get the desired explicit expression for $\widetilde{D}\gamma$ in terms of $f(x_0)$, $g(x_0)$, Df, and Dg. In particular, if $h \in \mathcal{A}(X)$ and $\gamma = [h]_{x_0}$ we can take $S = X$, $f = h$, and $g = 1$ and get $\widetilde{D}([h]_{x_0}) = Dh$, which proves that $D \mapsto \widetilde{D}$ is a right inverse for the canonical map $T_{X,x_0} \to TX_{x_0}$, so that the latter is surjective. To see that it is also injective we must show that any $D' \in T_{X,x_0}$ mapping onto D under the canonical map must agree with $\tilde{D}$. Now $[g]_{x_0}\gamma = [f]_{x_0}$ so that $Df = D'[f]_{x_0} = (D'[g]_{x_0})\gamma(x_0)+g(x_0)(D'\gamma) = (Dg)(f(x_0)/g(x_0))+g(x_0)(D'\gamma)$ and we easily derive $D'\gamma = (g(x_0)(Df)-f(x_0)(Dg))/g(x_0)^2 = \tilde{D}\gamma$. ■

1.10.12. Corollary. The canonical map $T^*X_{x_0} \to T^*_{X,x_0}$ is an isomorphism. The inverse map $\omega \mapsto \tilde{\omega}$ of $T^*_{X,x_0} \to T^*X_{x_0}$ can be characterized as follows: let $\omega = d\gamma$ for $\gamma \in \mathcal{O}_{X,x_0}$ and express γ in the form $[f]_{x_0}/[g]_{x_0}$ for some $f, g \in \mathcal{Q}(X)$ with $g(x_0) \neq 0$. Then:

$$\tilde{\omega} = (g(x_0)df_{x_0} - f(x_0)dg_{x_0})/g(x_0)^2.$$

Proof. Immediate from 1.10.11, 1.10.7, and 1.10.10. ■

1.10.13. Proposition. Let $\varphi : Y \to X$ be a morphism of ringed spaces. Let $y_0 \in Y$, $x_0 = \varphi(y_0)$ and suppose $\mathcal{O}_{\varphi,y_0} : \mathcal{O}_{X,x_0} \to \mathcal{O}_{Y,y_0}$ (namely $[f]_{x_0} \mapsto [f \circ \varphi]_{y_0}$, cf. 1.10.4) is a ring isomorphism. Then $T(\varphi)_{y_0}$: $TY_{y_0} \to TX_{x_0}$ and $T^*(\varphi)_{y_0} : T^*X_{x_0} \to T^*Y_{y_0}$ are linear isomorphisms.

Proof. Immediate from 1.10.11, 1.10.12, and 1.10.10. (Note that by 1.10.6, if $\mathcal{O}_{\varphi,y_0}$ is an isomorphism of rings, then $T_{\varphi,y_0} : T_{Y,y_0} \to T_{X,x_0}$ and $T^*_{\varphi,y_0} : T^*_{X,x_0}$ T^*_{Y,y_0} are linear isomorphisms.) ■

1.10.14. Remark. Let R be a local ring with maximal ideal m and let K denote the field R/m. If M is an R module then mM is a submodule. If $r \in R$ and $v \in M$ then

$$(r+m)(v+mM) = rv+mv+rmM+m^2M = rv+mM$$

so that the quotient module M/mM is a vector space over K (i.e., elements of R congruent modulo m operate identically on M/mM). Concerning this vector space we have the following important result.

Nakayama's Lemma. Suppose the R module M is finitely generated. Let $v_1, \ldots, v_n$ be elements of M and let $\bar{v}_i = (v_i+mM) \in M/mM$. Then $v_1, \ldots, v_n$ generate M (over R) if and only if $\bar{v}_1, \ldots, \bar{v}_n$ span M/mM (over K). In particular, the number of elements in any minimal set of generators for M over R is equal to $\dim_K(M/mM)$.

Proof. Consider first the case $mM = M$ or $\dim_K(M/mM) = 0$. In this case we must show that $M = 0$. Let $g_1, \ldots, g_n$ generate M. What we must show is that $g_1, \ldots, g_{n-1}$ also generate M. Now $g_n \in M = mM$ so $g_n = \alpha_1 u_1 + \ldots + \alpha_r u_r$ where $\alpha_i \in m$ and $u_i \in M$. Expressing each u_i in the form $\beta_1^i g_1 + \ldots + \beta_n^i g_n$ with $\beta_i \in R$ and collecting terms we have $g_n = \gamma_1 g_1 + \ldots + \gamma_n g_n$ where $\gamma_i = \alpha_1\beta_i^1 + \ldots + \alpha_r\beta_i^r \in m$ or $(1-\gamma_n)g_n = \gamma_1 g_1 + \ldots + \gamma_{n-1}g_{n-1}$. But since $1 \notin m$ and $\gamma_n \in m$ it follows $(1-\gamma_n) \notin m$. Since R is a local ring and m its maximal ideal, $1-\gamma_n$ has an inverse δ in R so that $g_n = \delta\gamma_1 g_1 + \ldots + \delta\gamma_{n-1}g_{n-1}$, and $g_1, \ldots, g_{n-1}$ generate. Now consider the general case, and assume that $\bar{v}_1, \ldots, \bar{v}_n$ span M/mM. Let N denote the submodule of M generated by $v_1, \ldots, v_n$. We must show that

$N = M$ or equivalently that $M/N = 0$. By the special case of the lemma just proved, it will suffice to show that $m(M/N) = (M/N)$. Given $v \in M$ we must show that $\tilde{v} = (v+N) \in M/N$ belongs to $m(M/N)$, i.e., that there exist $\alpha_1, \ldots, \alpha_r \in m$ and $u_1, \ldots, u_r \in M$ such that $v-(\alpha_1 u_1 + \ldots + \alpha_r u_r) \in N$. Equivalently we must show that there exists $n \in N$ such that $(v-n) \in mM$. Now by assumption $\bar{v} = (v+mM) \in M/mM$ is a linear combination of $\bar{v}_1, \ldots, \bar{v}_n$, which means $v-(r_1 v_1 + \ldots + r_n v_n)$ is in mM for some $r_1, \ldots, r_n \in R$. This completes the proof that if $\bar{v}_1, \ldots, \bar{v}_n$ span M/mM then $v_1, \ldots, v_n$ generate M, and the converse is trivial. ■

1.10.15. Remark. In applying the next proposition the following should be borne in mind. Let X be a ringed space over K, $x_0 \in X$, and let m denote the maximal ideal of $\mathcal{Q}(X)$ consisting of f vanishing at x_0. Then M_{X,x_0} consists of germs $\gamma = [f/g]_{x_0}$ where $f, g \in \mathcal{Q}(X)$, $f \in m$, and $g \notin m$. If $\gamma_1, \ldots, \gamma_n$ generate M_{X,x_0} as an ideal of $\mathcal{O}_{X,x_0}$, say $\gamma_i = [f_i/g_i]_{x_0}$, then so do $[f_1]_{x_0}, \ldots, [f_n]_{x_0}$, for if $\lambda_1, \ldots, \lambda_n \in \mathcal{O}_{X,x_0}$ then $\lambda_1\gamma_1 + \ldots + \lambda_n\gamma_n = \lambda_1[1/g_1]_{x_0}[f_1]_{x_0} + \ldots + \lambda_n[1/g_n]_{x_0}[f_n]_{x_0}$. Also if $f_1, \ldots, f_n$ generate m (as an ideal of $\mathcal{Q}(X)$) then $[f_1]_{x_0}, \ldots, [f_n]_{x_0}$ generate M_{X,x_0} as an ideal of $\mathcal{O}_{X,x_0}$. Indeed if $\gamma \in M_{X,x_0}$ say $\gamma = [f/g]_{x_0}$, then $f = h_1 f_1 + \ldots + h_n f_n$ with $h_i \in \mathcal{Q}(X)$ and so $\gamma = [h_1/g]_{x_0}[f_1]_{x_0} + \ldots + [h_n/g]_{x_0}[f_n]_{x_0}$. Thus for example if $\mathcal{Q}(X)$ is Noetherian, then M_{X,x_0} is always a finitely generated ideal in $\mathcal{O}_{X,x_0}$. In particular if $\mathcal{Q}(X)$ is finitely generated as an algebra, then m and hence M_{X,x_0} are finitely generated ideals (to see this directly note that if $u_1, \ldots, u_n$ generate $\mathcal{C}(X)$ as an algebra, then so clearly do $u_1 - u_1(x_0), \ldots, u_n - u_n(x_0)$, and the latter moreover generate m even as an algebra without identity, so a fortiori as an ideal of $\mathcal{C}(X)$).

1.10.16. Proposition. Let X be a ringed space over K, $x_0 \in X$ and assume that the maximal ideal M_{X,x_0} of $\mathcal{O}_{X,x_0}$ is finitely generated. Given $f_1, \ldots, f_n$ on $\mathcal{C}(X)$ vanishing at x_0, a necessary and sufficient condition that their germs $[f_1]_{x_0}, \ldots, [f_n]_{x_0}$ generate M_{X,x_0} is that $(df_1)_{x_0}, \ldots, (df_n)_{x_0}$ span $T^*X_{x_0}$, and the $(df_i)_{x_0}$ are a basis for $T^*X_{x_0}$ if and only if the $[f_i]_{x_0}$ are a minimal set of generators for $T^*X_{x_0}$.

Proof. In view of the canonical isomorphism of $T^*X_{x_0}$ with $T^*_{X,x_0} = M_{X,x_0}/M^2_{X,x_0}$ (see 1.10.9 and 1.10.12) this is just a special case of Nakayama's lemma (1.10.14) with $R = \mathcal{O}_{X,x_0}$, and $M = m = M_{X,x_0}$. ■

1.10.17. Remark. Let H denote the category of complex analytic manifolds and holomorphic mappings. We might try to make H into a ringed space category over $\mathbb{C}$ by taking as the structure ring of a complex analytic manifold M the algebra $H(M) = H(M, \mathbb{C})$ of holomorphic complex valued functions on M. It is clear that conditions (1a) to (1d) and (2) of 1.6.3 are satisfied, but unfortunately (1e) in general is not (if M is compact, for example the Riemann sphere, then the maximum modulus principle implies that $H(M)$ consists only of constant functions). Thus what turned out to be a phantom problem in the case of real C^{ω} manifolds (because of the beautiful and powerful theorems of H. Cartan and Grauert-Morrey, cf. sections 1.6.6, et seq.) is now real and unavoidable. The way out of this difficulty is by way of a generalization of the notion of ringed space to a considerably more sophisticated notion. This more general concept is also frequently called a "ringed space" but to avoid confusion we shall use the more precise term "local ringed space". While we shall not be concerned with this notion beyond this section it seems worthwhile to formulate the basic definitions, give references to more detailed treatments, and point out the relations with ringed spaces.

Returning to the category H above, let us associate to each open set U of a complex analytic manifold M the ring $\mathcal{O}_M(U)$ of holomorphic maps $f : U \to \mathbb{C}$, and whenever $U' \subseteq U$ let $j_{U',U} : \mathcal{O}_M(U) \to \mathcal{O}_M(U')$ denote the restriction homomorphism. We note that:

1) If $U'' \subseteq U' \subseteq U$ then $j_{U'',U} = j_{U'',U'} \circ j_{U'U}$.

2) If $\{U_\alpha\}_{\alpha \in A}$ is a collection of open sets in M with union U and $s_\alpha \in \mathcal{O}_M(U_\alpha)$ are such that whenever $\alpha, \beta \in A$, $j_{U_\alpha \cap U_\beta, U_\alpha}(s_\alpha) = j_{U_\alpha \cap U_\beta, U_\beta}(s_\beta)$ then there is a unique $s \in \mathcal{O}_M(U)$ such that $j_{U_\alpha, U}(s) = s_\alpha$.

We shall refer to these two properties as the "sheaf axioms". A <u>local ringed space</u> is a pair $(X, \mathcal{O}_X)$, where X is a topological space and $\mathcal{O}_X$ is a sheaf of rings on X; i.e., $\mathcal{O}_X$ assigns to each open set U of X a ring $\mathcal{O}_X(U)$ (<u>not necessarily</u> a ring of functions on U), and to each pair of open sets U, U' with $U' \subseteq U$ a "restriction" homomorphism $j_{U',U} : \mathcal{O}_X(U) \to \mathcal{O}_X(U')$ such that the above sheaf axioms are satisfied.

If $(X, \mathcal{A})$ is a ringed space we get an associated local ringed space $(X, \mathcal{O}_X)$ by giving X the Z-topology and for each Z-open set U of X letting $\mathcal{O}_X(U) = \mathcal{A}(U) = \{(f|U) | f \in \mathcal{A}\}$. The restriction homomorphisms are the obvious ones. The example of a compact, complex analytic local ringed space shows that not every local ringed space arises in this simple way.

If $(X, \mathcal{O}_X)$ and $(Y, \mathcal{O}_Y)$ are local ringed spaces, then a morphism $(X, \mathcal{O}_X) \to (Y, \mathcal{O}_Y)$ is a pair (ψ, θ) where $\psi : X \to Y$ is a continuous map and θ associates to each open set V of Y a homomorphism $\theta(V) : \mathcal{O}_Y(V) \to \mathcal{O}_X(\psi^{-1}(V))$ such that if $V' \subseteq V$ then we have commutativity in

$$\begin{array}{ccc} \mathcal{O}_Y(V) & \xrightarrow{\theta(V)} & \mathcal{O}_X(\psi^{-1}(V)) \\ \downarrow & & \downarrow \\ \mathcal{O}_Y(V') & \xrightarrow[\theta(V')]{} & \mathcal{O}_X(\psi^{-1}(V')). \end{array}$$

For example if $(X, \mathcal{A}(X))$ and $(Y, \mathcal{A}(Y))$ are ringed spaces over K and $\psi : X \to Y$ is a ringed space morphism, then we get a morphism (ψ, θ) of the associated local ringed spaces by defining $\theta(V)(g) = g \circ \psi$. Similarly if X and Y are complex manifolds and $\psi : X \to Y$ is a holomorphic map, then this same formula again defines a morphism of local ringed spaces. It is an easy exercise to see that in both of these cases all local ringed space morphisms in fact arise in this way.

A particularly important class of local ringed spaces are the so-called affine schemes $(X, \mathcal{O}_X)$ associated to commutative rings R. Given R let $X = \operatorname{Spec} p(R)$ denote the "prime spectrum of R", i.e., the set of all prime ideals of R. Given $f \in R$ we associate to f a function $\hat{f}$ on X, namely $\hat{f}(p)$ is the equivalence class of f modulo p. Thus $\hat{f}(p) = 0$ means $f \in p$. To each $f \in R$ we associate the "principal open set" $X_f \subseteq X$, $X_f = \{p \in X \mid \hat{f}(p) \neq 0\}$. We topologize X with the Zariski topology, namely the topology having the X_f as a base. (If $S \subseteq X$ then the closure of S is the set of p in X which include the intersection of all ideals of S, or equivalently the set of p in X such that every $\hat{f}$ vanishing on S vanishes at p.) The structure sheaf $\mathcal{O}_X$ assigns to a principal open set X_f the "ring of fractions" $R_f = \{g/f^n \mid g \in R,\ n \in \mathbb{Z}^+\}$. If $X_g \subseteq X_f$ then $\mathcal{O}_X$ assigns the natural homomorphism $R_f \to R_g$ (namely $g^n = sf$ for some $n \in \mathbb{Z}^+$ and $s \in R$ and we map a/f^m in R_f to as^m/g^{mn} in R_g). It is not hard to see that these assignments characterize a unique sheaf $\mathcal{O}_X$ on X. A local ringed space $(X, \mathcal{O}_X)$ is called a <u>prescheme</u> if each point of X has a neighborhood isomorphic (as a local ringed space) to such an affine scheme. For further details on the above definitions and an introduction to the theory of preschemes and schemes, the reader is referred to one of the following: Borel [4], Hartshorne [13], Macdonald [20], Mumford [23], Shaferevich [27].

For the most general theory of algebraic geometry, there is now a reasonably universal consensus that the theory of schemes is an ideal setting. For the applications that seem to arise in differential topology however, particularly in investigating the relationship between smooth manifolds and smooth real algebraic varieties, the much simpler theory of ringed spaces seems to be exactly what is needed.

1.11. Equivalence Relations.

Let X be a set and let $\mathcal{E} \subseteq X \times X$ be an equivalence relation in X. We will sometimes write $x_1 \mathcal{E} x_2$ to mean $(x_1, x_2) \in \mathcal{E}$. If $x_0 \in \mathcal{E}$ then $[x_0]_{\mathcal{E}} = \{x \in X \mid x \mathcal{E} x_0\}$ denotes its equivalence class modulo $\mathcal{E}$. We write $X/\mathcal{E}$ for the set of equivalence classes of X modulo $\mathcal{E}$ and $\Pi_{\mathcal{E}} : X \to X/\mathcal{E}$ for the canonical map, so $\Pi_{\mathcal{E}}(x) = [x]_{\mathcal{E}}$. If $S \subseteq X$ then its saturation relative to $\mathcal{E}$ is $\bigcup_{s \in S} [s]_{\mathcal{E}}$, or equivalently $\Pi_{\mathcal{E}}^{-1}(\Pi_{\mathcal{E}}(S))$, and S is called $\mathcal{E}$-saturated if it is equal to its saturation, i.e., if $x_0 \in S$ and $x \mathcal{E} x_0$ implies $x \in S$.

Given any function f mapping X to some set S we get an equivalence relation $\mathcal{E}_f$ in X by defining $(x_1, x_2) \in \mathcal{E}_f$ if and only if $f(x_1) = f(x_2)$, so the equivalence classes of $\mathcal{E}_f$ are the "fibers" of f, i.e., the inverse images $f^{-1}(s)$ where $s \in \operatorname{im}(f)$. (Note that every equivalence relation $\mathcal{E}$ arises in this way, namely by taking $f = \Pi_{\mathcal{E}}$.) We call f <u>consistent with</u> $\mathcal{E}$ if "$\mathcal{E}$ implies $\mathcal{E}_f$", i.e., if $x_1 \mathcal{E} x_2$ implies $f(x_1) = f(x_2)$. In this case f is constant on each equivalence class modulo $\mathcal{E}$ and induces a uniquely determined map $\tilde{f} : X/\mathcal{E} \to S$ satisfying $\tilde{f} \circ \Pi_{\mathcal{E}} = f$.

Now suppose X is the underlying set of some object in a category $\mathcal{C}$ of structured sets. A structure for $X/\mathcal{E}$ as an object of $\mathcal{C}$ is called a <u>quotient structure for</u> X <u>modulo</u> $\mathcal{E}$ (relative to $\mathcal{C}$) if $\Pi_{\mathcal{E}} : X \to X/\mathcal{E}$ is a morphism of $\mathcal{C}$ and for every morphism $f : X \to S$ of $\mathcal{C}$ which is consistent with $\mathcal{E}$, the induced map $\tilde{f} : X/\mathcal{E} \to S$ is also a morphism of $\mathcal{C}$. This structure, if it exists, is clearly unique (for taking $f = \Pi_{\mathcal{E}}$, it follows that the identity map $X/\xi \to X/\xi$ is a morphism from any one such structure for X/ξ to any other). For a given $\mathcal{C}$ some interesting questions to answer are: under what conditions on ξ does a quotient structure for X/ξ exist, what important special properties of objects of $\mathcal{C}$ are preserved in the passage from X to X/ξ, if we have a weakening of structure functor $\mathcal{C} \rightsquigarrow \mathcal{C}'$ where $\mathcal{C}'$ is some other category of structured sets, does this carry quotients in the sense of $\mathcal{C}$ to quotients in the sense of $\mathcal{C}'$?

Before considering quotients for the category of ringed spaces let us consider a few better known cases. First consider the category TOP of topological spaces and continuous maps. It is well known that quotients always exist in this case. The quotient topology for X/ξ is simply the strongest topology for X/ξ such that $\Pi_\xi : X \to X/\xi$ is continuous, i.e., a subset S of X/ξ is open (closed) if and only if its inverse image $\Pi_\xi^{-1}(S)$ (the union of the classes in S) is open (closed) in X. If instead of TOP we consider the full subcategory T_1 of T_1 spaces, then X/ξ exists if and only if each equivalence class of X modulo ξ is closed in X; for the full subcategory T_2 of Hausdorff spaces X/ξ exists if and only if ξ is closed in $X \times X$ (with its product topology) and in both cases the quotients are the same in the subcategory as in TOP. Any property of topological spaces preserved by surjective continuous maps (e.g., compactness, connectedness, arcwise connectedness) is preserved in going from X to X/ξ. (The analogous remark of course applies quite generally to any category $\mathcal{C}$ of structured sets: any property preserved by surjective morphisms is preserved under taking quotients.)

Next consider the category C^k of C^k manifolds ($k = 1, 2, \ldots, \infty, \omega$). In this case $X/\mathcal{E}$ exists if and only if the following two conditions hold

1) $\mathcal{E}$ is a regularly embedded closed C^k submanifold of $X \times X$.

2) The natural projection of $\mathcal{E} \subseteq X \times X$ onto either (and hence by symmetry both) factors is a submersion.

It is relatively trivial to see that these conditions are necessary for a quotient C^k structure to exist on $X/\mathcal{E}$. However, there sufficiency is decidedly non-trivial; cf. Theorem 2, p. 3.27 of [4].

We now investigate the conditions under which an equivalence relation $\mathcal{E}$ on a ringed space X (over K) will admit a quotient ringed space structure on $X/\mathcal{E}$.

1.11.1. <u>Definition</u>. Let X be a ringed space over K and let $\mathcal{E}$ be an equivalence relation in X. Let $\mathcal{a}(X)^{\mathcal{E}}$ denote the subalgebra of $\mathcal{a}(X)$ consisting of functions $f : X \to K$ consistent with $\mathcal{E}$ (i.e., constant on each equivalence class modulo $\mathcal{E}$). For each $f \in \mathcal{a}(X)^{\mathcal{E}}$ let $\tilde{f} : X/\mathcal{E} \to K$ denote the unique function such that $\tilde{f} \circ \prod_{\mathcal{E}} = f$ and let $\mathcal{a}(X/\mathcal{E})$ denote the algebra of K-valued functions on $X/\mathcal{E}$ of the form $\tilde{f}$ for some $f \in \mathcal{a}(X)^{\mathcal{E}}$, so that $\tilde{f} \mapsto \tilde{f} \circ \prod_{\mathcal{E}}$ is an algebra isomorphism $\prod_{\mathcal{E}}^{*} : \mathcal{a}(X/\mathcal{E}) \to \mathcal{a}(X)^{\mathcal{E}}$. We call $\mathcal{E}$ a <u>ringed equivalence relation in</u> X if $\mathcal{a}(X/\mathcal{E})$ separates points of $X/\mathcal{E}$, or equivalently if $\mathcal{a}(X)^{\mathcal{E}}$ separates equivalence classes of $\mathcal{E}$ (i.e., given $x_1, x_2 \in X$ with $(x_1, x_2) \notin \mathcal{E}$ there exists $f \in \mathcal{a}(X)^{\mathcal{E}}$ such that $f(x_1) \neq f(x_2)$). In this case we will regard $X/\mathcal{E}$ as a ringed space over K with structure ring $\mathcal{a}(X/\mathcal{E})$, and we note that it is immediate from the definition that $\prod_{\mathcal{E}} : X \to X/\mathcal{E}$ is a morphism of ringed spaces.

1.11.2. Remark. Let Z denote $X/\mathcal{E}$ with any structure ring $\mathcal{Q}(Z)$ making $\prod_{\mathcal{E}} : X \to Z$ a ringed space morphism. If $g \in \mathcal{Q}(X)$ then $f = g \circ \prod_{\mathcal{E}}$ is in $\mathcal{Q}(X)$, and since f is clearly consistent with $\mathcal{E}$, $f \in \mathcal{Q}(X)^{\mathcal{E}}$ which means $g \in \mathcal{Q}(X/\mathcal{E})$. Thus $\mathcal{Q}(Z) \subseteq \mathcal{Q}(X/\mathcal{E})$. Since $\mathcal{Q}(Z)$ separates points, so does $\mathcal{Q}(X/\mathcal{E})$, hence $\mathcal{E}$ is a ringed equivalence relation on X. Moreover the identity map $i : X/\mathcal{E} \to Z$ is a ringed space morphism. Suppose Z is the quotient of X modulo $\mathcal{E}$ in the category of ringed spaces. Since $\prod_{\mathcal{E}} : X \to X/\mathcal{E}$ is trivially consistent with $\mathcal{E}$, it follows from the definition of quotient object in a category that the identity map $Z \to X/\mathcal{E}$ is a morphism of ringed spaces, i.e., $\mathcal{Q}(X/\mathcal{E}) \subseteq \mathcal{Q}(Z)$. Thus $\mathcal{Q}(X/\mathcal{E}) = \mathcal{Q}(Z)$ and $Z = X/\mathcal{E}$. In other words what we have shown is that if X admits a quotient modulo $\mathcal{E}$ in the category of ringed spaces then $\mathcal{E}$ is a ringed equivalence relation and the quotient object has as its structure ring the ring $\mathcal{Q}(X/\mathcal{E})$ defined in 1.11.1.

1.11.3. Proposition. If $\mathcal{E}$ is an equivalence relation on a ringed space X, then a necessary and sufficient condition for $X/\mathcal{E}$ to admit the structure of quotient object for X modulo $\mathcal{E}$ in the category of ringed spaces is that $\mathcal{E}$ be a ringed equivalence relation in X. In this case the structure ring of the quotient object is the ring $\mathcal{Q}(X/\mathcal{E})$ defined in 1.1.1.

Proof. Necessity, and the fact that $\mathcal{Q}(X/\mathcal{E})$ is the only possible choice of structure ring for the quotient object we have just seen in 1.11.2. So assume $\mathcal{E}$ is a ringed equivalence relation in X. Let $h : X \to S$ be a morphism of ringed spaces consistent with $\mathcal{E}$ and let $\tilde{h} : X/\mathcal{E} \to S$ be the induced map, so $h = \tilde{h} \circ \Pi_{\mathcal{E}}$. We must show that if $g \in \mathcal{Q}(S)$ then $g \circ \tilde{h} \in \mathcal{Q}(X/\mathcal{E})$. Now $f = g \circ h \in \mathcal{Q}(X)$ and since $x_1 \mathcal{E} x_2$ implies $h(x_1) = h(x_2)$ it follows that $x_1 \mathcal{E} x_2$ also implies $f(x_1) = f(x_2)$, so f is consistent with $\mathcal{E}$, i.e., there is a unique map $\tilde{f} : X/\mathcal{E} \to K$ such that $\tilde{f} \circ \Pi_{\mathcal{E}} = f$. Moreover since $f \in \mathcal{Q}(X)$ we have $f \in \mathcal{Q}(X)^{\mathcal{E}}$ and hence $\tilde{f} \in \mathcal{Q}(X/\mathcal{E})$. Now $f = g \circ h = g \circ \tilde{h} \circ \Pi_{\mathcal{E}}$ so that $\tilde{f} = g \circ h$. ∎

1.11.4. <u>Proposition</u>. Let $\mathcal{E}$ be an equivalence relation on a ringed space X. Let $\mathcal{J}_{\mathcal{E}}$ denote the ideal in $\mathcal{Q}(X \times X) = \mathcal{Q}(X) \otimes \mathcal{Q}(X)$ generated by elements of the form $f \otimes 1 - 1 \otimes f$ where $f \in \mathcal{Q}(X)^{\mathcal{E}}$. Then $\mathcal{E} \subseteq V(\mathcal{J}_{\mathcal{E}})$, and $\mathcal{E}$ is a ringed equivalence relation in X if and only if $\mathcal{E} = V(\mathcal{J}_{\mathcal{E}})$.

<u>Proof.</u> If $x \mathcal{E} y$ and $f \in \mathcal{Q}(X)^{\mathcal{E}}$ then $(f \otimes 1 - 1 \otimes f)(x, y) = f(x) - f(y) = 0$ which proves $\mathcal{E} \subseteq V(\mathcal{J}_{\mathcal{E}})$. If $(x, y) \notin \mathcal{E}$ then a necessary and sufficient condition that an element $f \in \mathcal{Q}(X/\mathcal{E})$ separate $[x]_{\mathcal{E}}$ from $[y]_{\mathcal{E}}$ is that $(f \otimes 1 - 1 \otimes f)(x, y) \neq 0$, where $f = \Pi_{\mathcal{E}} \circ \tilde{f} \in \mathcal{Q}(X)^{\mathcal{E}}$. Thus $\mathcal{Q}(X/\mathcal{E})$ separates points of $X/\mathcal{E}$ (i.e., $\mathcal{E}$ is a ringed equivalence relation) if and only if points of $X \times X$ not in $\mathcal{E}$ are not in $V(\mathcal{J}_{\mathcal{E}})$, i.e., if and only if $V(\mathcal{J}_{\mathcal{E}}) \subseteq \mathcal{E}$. ∎

1.11.5. <u>Corollary</u>. If $\mathcal{E}$ is a ringed equivalence relation in X then $\mathcal{E}$ is Z-closed in $X \times X$.

1.11.6. <u>Corollary</u>. If $\mathcal{E}$ is a ringed equivalence relation in X then each equivalence class of X modulo $\mathcal{E}$ is Z-closed in X.

<u>Proof.</u> The equivalence classes of $\mathcal{E}$ are inverse images of points under

$\prod_{\mathcal{E}} : X \to X/\mathcal{E}$, and since $\prod_{\mathcal{E}}$ is a morphism and hence Z-continuous, the result follows. However, it is worth noting that this also follows from the fact that $\mathcal{E}$ is Z-closed in $X \times X$. Indeed, if $x_0 \in X$, then $j_{x_0} : X \to X \times X$, $x \mapsto (x, x_0)$ is a ringed space morphism (and in fact maps X isomorphically onto its image, $X \times \{x_0\}$). Moreover $[x_0]_{\mathcal{E}} = j^{-1}(\mathcal{E})$. ■

1.11.7. Proposition. Let $\mathcal{E}$ be an equivalence relation on a ringed space X. Let F be an $\mathcal{E}$ saturated subset of X and consider the following condition:

(*) Given $x \in X - F$ there exists $f \in \mathcal{Q}(X)^{\mathcal{E}}$ such that $f|F = 0$ and $f(x) \neq 0$.

A necessary and sufficient condition that $\mathcal{E}$ be a ringed equivalence relation on X is that (*) holds whenever F is an equivalence class of X modulo $\mathcal{E}$. In this case a necessary and sufficient condition that the Z-topology for $X/\mathcal{E}$ be the quotient topology of the Z-topology for X is that (*) hold for all $\mathcal{E}$-saturated Z-closed subsets F of X.

Proof. The first conclusion is just a trivial restatement of what it means for $\mathcal{E}$ to be a ringed equivalence relation. Given $\tilde{F} \subseteq X/\mathcal{E}$ let $F = \prod_{\mathcal{E}}^{-1}(\tilde{F})$. Then (*) says that given $\tilde{x} = \prod_{\mathcal{E}}(x)$ not in $\tilde{F}$ there exists $\tilde{f} \in \mathcal{Q}(X/\mathcal{E})$ such that $\tilde{f}|\tilde{F} = 0$ and $\tilde{f}(\tilde{x}) \neq 0$, i.e., that $\tilde{F}$ is Z-closed in $X/\mathcal{E}$. Thus (*) holds for all $\mathcal{E}$-saturated Z-closed subsets of X if and only if $\prod_{\mathcal{E}}^{-1}(\tilde{F})$ Z-closed in X implies $\tilde{F}$ is Z-closed in $X/\mathcal{E}$, i.e., if and only if the Z-topology for $X/\mathcal{E}$ is the quotient of the Z-topology for X. ■

1.11.8. Proposition. If S_1 and S_2 are subsets of a ringed space X over

K, then the following are equivalent:

1) No K homomorphism of $\mathcal{C}(X)$ onto an extension field of K vanishes on both $\mathcal{I}(S_1)$ and $\mathcal{I}(S_2)$.

2) No maximal ideal of $\mathcal{C}(X)$ includes both $\mathcal{I}(S_1)$ and $\mathcal{I}(S_2)$.

3) $\mathcal{C}(X) = \mathcal{I}(S_1) + \mathcal{I}(S_2)$.

4) The restriction map $f \mapsto f|S_2$ maps $\mathcal{I}(S_1)$ onto $\mathcal{C}(S_2)$.

5) There exists $f \in \mathcal{C}(X)$ such that $f|S_1 = 0$ and $f|S_2 = 1$.

Proof. 1) <=> 2) <=> 3) $\Rightarrow$ 4) $\Rightarrow$ 5) are all easy (for 2) $\Rightarrow$ 3) use Zorn's lemma). Given 5) we have $1 = f + (1-f)$ where $f \in \mathcal{I}(S_1)$ and $(1-f) \in \mathcal{I}(S_2)$ and since $\mathcal{I}(S_1) + \mathcal{I}(S_2)$ is an ideal, 3) follows. ■

1.11.9. Definition. Subsets S_1 and S_2 of a ringed space X over K will be called strongly separated if they satisfy any one and hence all of the conditions of 1.11.8. A subset S of X will be called absolutely closed in X if S is strongly separated from every Z-closed subset of X which is disjoint from S.

1.11.10. Proposition. Let X be a ringed space. An absolutely closed subset of X is Z-closed. Each point of X is absolutely closed in X and finite unions of absolutely closed subsets of X are absolutely closed, so every finite subset of X is absolutely closed.

Proof. It is clear that a subset S of X is Z-closed if and only if S is strongly separated from each one point set $\{p\}$ such that $p \notin S$. From this it follows both that points are absolutely closed and that absolutely closed sets are Z-closed. It is also clear that if S is strongly separated from each of S_1 and S_2 then it is strongly separated from $S_1 \cup S_2$ (if $f_i \in \mathcal{A}(X)$ and $f_i | S_i = 0$, $f_i | S = 1$, then $f = f_1 f_2$ vanishes on $S_1 \cup S_2$ and is 1 on S). It follows that if S_1 and S_2 are absolutely closed so is $S_1 \cup S_2$. ∎

1.11.11. Example. Let X be a completely regular topological space, considered as a ringed space over $\mathbb{R}$ with structure ring the ring $C_B(X)$ of bounded, continuous, real valued functions on X (cf. 1.6.2). We recall that the topology for X is also the Z-topology for $(X, C_B(X))$. By Urysohn's Lemma if X is normal, then any two disjoint closed sets in X are strongly separated and hence every closed subset of X is absolutely closed. If C is a compact subset of X and S is a disjoint closed set, then applying the above result to C and the closure of S in the Čech compactification of X we see that C and S are strongly separated; hence every compact subset of X is absolutely closed even if X is not necessarily normal.

1.11.12. Proposition. Let X be a complete ringed space over an algebraically closed field K and assume $\mathcal{A}(X)$ is finitely generated as a K-algebra. Then any two disjoint Z-closed subsets of X are strongly separated, and hence every Z-closed subset of X is absolutely closed in X.

Proof. Let S_1 and S_2 be disjoint Z-closed subsets of X and let φ be a K-homomorphism of $\mathcal{A}(X)$ onto an extension field F of K which vanishes on $\mathcal{I}(S_1)$ and $\mathcal{I}(S_2)$. To say that $\mathcal{A}(X)$ is finitely generated as a K algebra means it is the homomorphic image of some polynomial algebra $K[X_1, \ldots, X_n]$, and since F is the homomorphic image of $\mathcal{A}(X)$ the same is true of F. As we shall prove later (Nullstellensatz Lemma, 2.2.10) this implies that F is algebraic over K, and hence, since K is algebraically closed, that $F = K$ so that $\varphi \in \mathcal{C}(X)^{\wedge}$. Since X is complete there exists $x \in X$ such that $\varphi(f) = f(x)$ for all $f \in \mathcal{C}(X)$. If $f \in \mathcal{I}(S_1)$ then $f(x) = \varphi(f) = 0$ so $x \in V(\mathcal{I}(S_1))$ and since S_1 is closed in X, $x \in S_1$. Similarly $x \in S_2$, contradicting the fact that S_1 and S_2 are disjoint. ∎

1.11.13. Definition. If $\mathcal{E}$ is an equivalence relation on a ringed space X, a K linear map $A : \mathcal{C}(X) \to \mathcal{C}(X)$ is called an averaging map for $\mathcal{E}$ if:

(1) $A(1) = 1$.

(2) If $f \in \mathcal{C}(X)$ vanishes on an $\mathcal{E}$ equivalence class, then Af also vanishes on this class.

1.11.14. Proposition. If $\mathcal{E}$ is an equivalence relation on the ringed space X and $A : \mathcal{C}(X) \to \mathcal{C}(X)^{\mathcal{E}}$ is an averaging map for $\mathcal{E}$ then:

1) If f is constant on an $\mathcal{E}$ equivalence class, then Af agrees with f on this class.

2) A is a projection of $\mathcal{C}(X)$ onto $\mathcal{C}(X)^{\mathcal{E}}$.

Proof. Suppose f has the constant value c on a class $[x]_{\mathcal{E}}$. Then $g = f - c1$ vanishes on $[x]_{\mathcal{E}}$, so $A(f) - c1 = A(f - c1) = A(g)$ vanishes on $[x]_{\mathcal{E}}$, i.e., $A(f)$ equals c on $[x]_{\mathcal{E}}$, proving 1). If $f \in \mathcal{C}(X)^{\mathcal{E}}$ then f is constant on every equivalence class of X modulo $\mathcal{E}$ so that by 1) Af agrees with f on all equivalence classes of $\mathcal{E}$, i.e., $Af = f$. This proves 2). ∎

1.11.15. Proposition. Let X be a ringed space and let $\mathcal{E}$ be an equivalence relation in X such that

1) every equivalence class of X modulo $\mathcal{E}$ is absolutely closed in X;

2) $\mathcal{E}$ admits an averaging operator $A : \mathcal{C}(X) \to \mathcal{C}(X)^{\mathcal{E}}$.

Then $\mathcal{E}$ is ringed equivalence relation in X and the Z-topology for $X/\mathcal{E}$ is the quotient of the Z-topology for X.

Proof. Let F be an $\mathcal{E}$-saturated Z-closed subset of X and let $x \in X-F$. Since F is $\mathcal{E}$-saturated, $[x]_{\mathcal{E}}$ is disjoint from F and since $[x]_{\mathcal{E}}$ is absolutely closed in X it is strongly separated from F, i.e., there exists $f \in \mathcal{C}(X)$ such that $f|F = 0$ and f equals 1 on $[x]_{\mathcal{E}}$. Then by 1.11.14, since F is $\mathcal{E}$-saturated, $(Af)|F = 0$ and Af equals 1 on $[x]_{\mathcal{E}}$. Since $Af \in \mathcal{C}(X)^{\mathcal{E}}$ it follows that (*) of 1.11.7 holds for all $\mathcal{E}$-saturated Z-closed subset of X. By 1) every equivalence class of $\mathcal{E}$ is Z-closed so (*) of 1.11.7 holds in particular when F is a single such class and the conclusions now follow from 1.11.7. ■

1.11.16. Definition. Let X be a ringed space and let Γ be a group of (ringed space) automorphisms of X (i.e., a subgroup of the group of ringed space isomorphisms $g : X \to X$). We define an equivalence relation $\mathcal{E}_\Gamma$ in X by $\mathcal{E}_\Gamma = \{(x, \gamma x) | x \in X, \gamma \in \Gamma\}$. We write Γx for the equivalence class of x modulo $\mathcal{E}_\Gamma$ and we call Γx the orbit (or Γ-orbit) of x. We write X/Γ for $X/\mathcal{E}_\Gamma$ and call it the orbit space of Γ, and we write $\Pi_\Gamma : X \to X/\Gamma$ for the canonical "orbit map", $x \mapsto \Gamma x$. Finally we write $\mathcal{C}(X)^\Gamma$ for $\mathcal{C}(X)^{\mathcal{E}_\Gamma}$, i.e., $\mathcal{C}(X)^\Gamma$ is the set of all $f \in \mathcal{C}(X)$ such that $f(gx) = f(x)$ for all $x \in X$ and $\gamma \in \Gamma$ or equivalently the set of all $f \in \mathcal{C}(X)$ such that $g^* f = f$ for all $\gamma \in \Gamma$.

1.11.17. <u>Remark.</u> Let X be a ringed space and Γ a group of automorphisms of X. If $\mathcal{O}$ is Z-open in X then its saturation, $\Pi_\Gamma^{-1}(\Pi_\Gamma(\mathcal{O}))$, is clearly equal to $\bigcup_{\gamma\in\Gamma} \gamma(\mathcal{O})$, and since $\gamma(\mathcal{O})$ is Z-open (because $\gamma : X \to X$ is an isomorphism of ringed spaces and hence a homeomorphism with respect to the Z-topology) it follows that $\Pi_\Gamma^{-1}(\Pi_\Gamma(\mathcal{O}))$ is Z-open in X. Recalling the definition of the quotient topology for X/Γ this says that $\Pi_\Gamma : X \to X/\Gamma$ is an open map when X is given its Z-topology and X/Γ the corresponding quotient topology.

1.11.18. <u>Remark.</u> Let Γ be a <u>finite</u> group of automorphisms of the ringed space X, say of order $|\Gamma|$. We note that $\gamma \mapsto (\gamma^{-1})^*$ is a homomorphism of Γ into the group of linear automorphisms of $\mathcal{Q}(X)$ (i.e., a representation of Γ in $\mathcal{Q}(X)$) or in other words that $\mathcal{Q}(X)$ is a Γ-module (over K). We write γf for $(\gamma^{-1})^* f$ if $\gamma \in \Gamma$ and $f \in \mathcal{Q}(X)$. It is trivial that $\sum_{\gamma\in\Gamma} \gamma f \in \mathcal{Q}(X)^\Gamma$ for all $f \in \mathcal{Q}(X)$. If $f \in \mathcal{Q}(X)^\Gamma$ then $\sum_{\gamma\in\Gamma} \gamma f = |\Gamma| f$; hence if the characteristic of K does not divide $|\Gamma|$ then

$$Af = |\Gamma|^{-1} \sum_{\gamma\in\Gamma} \gamma f$$

defines a map $A : \mathcal{Q}(X) \to \mathcal{Q}(X)^\Gamma$ which is trivially seen to be an averaging map for $\mathcal{E}_\Gamma$. Now if $\gamma \in \Gamma$ and $f \in \mathcal{Q}(X)$ then clearly $A\gamma f = Af = \gamma Af$, i.e., $A : \mathcal{Q}(X) \to \mathcal{Q}(X)$ is Γ-equivariant. Conversely if $T : \mathcal{Q}(X) \to \mathcal{Q}(X)$ is Γ-equivariant and is an averaging map for $\mathcal{E}_\Gamma$, or even if T is a linear projection of $\mathcal{Q}(X)$ onto $\mathcal{Q}(X)^\Gamma$, then T must equal A. For $T\gamma f = \gamma Tf = Tf$, hence summing over all $\gamma \in \Gamma$ and dividing by $|\Gamma|$ gives (using the linearity of T) $TAf = Tf$

and since $Af \in \mathcal{A}(X)^{\Gamma}$ and T is a projection onto $\mathcal{A}(X)^{\Gamma}$, $Tf = TAf = Af$. Thus A is the unique Γ-equivariant averaging map for $\mathcal{E}_{\Gamma}$.

1.11.19. Theorem. Let X be a ringed space over K and let Γ be a finite group of automorphisms of X. If the characteristic of K does not divide the order $|\Gamma|$ of Γ then $\mathcal{E}_{\Gamma} = \{(x, \gamma x) \mid x \in X, \gamma \in \Gamma\}$ is a ringed equivalence relation on X. Moreover the Z-topology for X/Γ is the quotient of the Z-topology for X, and $\Pi_{\Gamma} : X \to X/\Gamma$ is an open map.

Proof. Since all orbits Γx are finite (and in fact consist of at most $|\Gamma|$ points) they are absolutely closed in X by 1.11.10. Thus by 1.11.15 and 1.11.17 it will suffice to have an averaging operator $A : \mathcal{A}(X) \to \mathcal{A}(X)^{\mathcal{E}_{\Gamma}} = \mathcal{A}(X)^{\Gamma}$. But this exists by 1.11.18. ∎

1.11.20. Proposition. Let X be a complete ringed space, $\mathcal{E}$ a ringed equivalence relation in X and $j : \mathcal{A}(X)^{\mathcal{E}} \to \mathcal{A}(X)$ the inclusion homomorphism. A necessary and sufficient condition for $X/\mathcal{E}$ to be complete is that $\hat{j} : \mathcal{A}(X)^{\wedge} \to (\mathcal{A}(X)^{\mathcal{E}})^{\wedge}$ be surjective.

Proof. Let c denote the isomorphism $f \mapsto f \circ \Pi_{\mathcal{E}}$ of $\mathcal{A}(X/\mathcal{E})$ with $\mathcal{A}(X)^{\mathcal{E}}$. Recalling that $X/\mathcal{E}$ is complete if and only if $Ev : X/\mathcal{E} \to \mathcal{A}(X/\mathcal{E})^{\wedge}$ is surjective, the result follows from the obvious commutativity of

$$\begin{array}{ccccc}
X & & \xrightarrow{\Pi_{\mathcal{E}}\ \text{(surjective)}} & & X/\mathcal{E} \\
{\scriptstyle Ev}\downarrow \ \text{(surjective)} & & & & \downarrow{\scriptstyle Ev} \\
\mathcal{A}(X)^{\wedge} & \xrightarrow[\hat{j}]{} & (\mathcal{A}(X)^{\mathcal{E}})^{\wedge} & \xrightarrow[\hat{c}]{\text{(bijective)}} & \mathcal{A}(X/\mathcal{E})^{\wedge}.
\end{array}$$

∎

1.11.21. Example. Let K be an infinite field considered as a complete ringed space with structure ring $\mathcal{P}(K) = K[X]$ (cf. 1.2.11). Note that by 1.7.18 K is irreducible. Let $\gamma : K \to K$ be the automorphism of order 2 (involution) $\alpha \mapsto -\alpha$, so that $\Gamma = \{e, \gamma\}$ is a finite group of automorphisms of K. Clearly $K[X]^{\Gamma} = K[X^2]$. We have an isomorphism $\tau : K[X^2] \approx K[X]$, $f(X^2) \mapsto f(X)$ which induces a bijection $\hat{\tau} : K[X]^{\wedge} \approx K[X^2]^{\wedge}$. The squaring map $S : K \to K$, $\alpha \mapsto \alpha^2$ is a morphism and $S^* : K[X] \to K[X]$ is $f(X) \mapsto f(X^2)$, so that $S^* \circ \tau : K[X^2] \to K[X]$ is just the inclusion $j : K[X^2] \to K[X]$, $\hat{j} = \hat{\tau} \circ S$. Thus $\hat{j} : K[X]^{\wedge} \to (K[X]^{\Gamma})^{\wedge}$ is surjective if and only if $S : K \to K$ is surjective, i.e., if and only if each element of K has a square root in K. Thus by 1.11.20 this is also the condition that K/Γ is complete and we see that even in the case of a very good equivalence relation of the type described in 1.11.19 the quotient of a complete space need not be complete.

1.11.22. Proposition. Let X be a ringed space over K and for each subalgebra $B \subseteq \mathcal{A}(X)$ let $\mathcal{E}_B$ denote the equivalence relation in X:

$$\mathcal{E}_B = \{(x_1, x_2) \in X \times X \mid f(x_1) = f(x_2) \text{ for all } f \in B\}.$$

The equivalence relations we get in this way are exactly the ringed equivalence relations in X. In general if $\mathcal{E}$ is an equivalence relation in X then $\mathcal{E} \subseteq \mathcal{E}_{\mathcal{A}(X)^{\mathcal{E}}}$ with equality if and only if $\mathcal{E}$ is a ringed equivalence relation. In general if B is a subalgebra of $\mathcal{A}(X)$ then $B \subseteq \mathcal{A}(X)^{\mathcal{E}_B}$ and equality holds if and only if B is of the form $\mathcal{A}(X)^{\mathcal{E}}$ (in which case we can take $\mathcal{E} = \mathcal{E}_B$ so $\mathcal{E}$ is a ringed equivalence relation on X).

Proof. Let P denote the set of all equivalence relations in X and let Q denote the set of all subalgebras of $\mathcal{C}(X)$, considered as partially ordered sets under inclusion. We have maps $\varphi : P \to Q$ and $\psi : Q \to P$ given by $\mathcal{E} \mapsto \mathcal{C}(X)^{\mathcal{E}}$ and $B \mapsto \mathcal{E}_B$ respectively. We leave to the reader the straightforward verification that the conditions for a Galois connection are satisfied (cf. 1. 3. 8). Then everything follows from the theorem of 1. 3. 8 except the fact that $\mathcal{E}_{\mathcal{C}(X)^{\mathcal{E}}} \subseteq \mathcal{E}$ if and only if $\mathcal{E}$ is a ringed equivalence relation on X. But if $x_1, x_2 \in X$ and $(x_1, x_2) \notin \mathcal{E}$ then a necessary and sufficient condition that $[x_1]_{\mathcal{E}}$ and $[x_2]_{\mathcal{E}}$ can be separated by an element of $\mathcal{C}(X/\mathcal{E})$ is that there exist f in $\mathcal{C}(X)^{\mathcal{E}}$ such that $f(x_1) \neq f(x_2)$, i.e., that $(x_1, x_2) \notin \mathcal{E}_{\mathcal{C}(X)^{\mathcal{E}}}$. ■

1. 11. 23. Remark. Let $f : X \to Y$ be a ringed space morphism and let $\mathcal{E}_f = \{(x_1, x_2) \in X \times X \mid f(x_1) = f(x_2)\}$. If $B = f^*(\mathcal{C}(Y)) = \{g \circ f \mid g \in \mathcal{C}(Y)\}$, then trivially $\mathcal{E}_f \subseteq \mathcal{E}_B$, and since $\mathcal{C}(Y)$ separates points of Y it follows that in fact $\mathcal{E}_f = \mathcal{E}_B$. Thus by 1. 11. 23 $\mathcal{E}_f$ is a ringed equivalence relation in X and moreover $f^*(\mathcal{C}(Y)) \subseteq \mathcal{C}(X)^{\mathcal{E}_f}$.

1. 11. 24. Definition. A morphism of ringed spaces $f : X \to Y$ is called a weakening morphism if it is bijective.

1. 11. 25. Remark. If X is any ringed space and $\widetilde{X}$ is the regularization of X then the identity map $\widetilde{X} \to X$ is a weakening morphism. In general any injective map $f : X \to Y$ can be considered as a weakening morphism $f : X \to f(X)$ where we regard $f(X)$ as a ringed subspace of Y. We note that

if $f : X \to Y$ is a weakening morphism and X is complete, Y need not be complete. For example, let V be a complete ringed space over K (e.g., K itself with the structure ring $K[X]$, cf. 1.2.11). Let $\mathcal{O}$ be a non-Z-closed subset of V of the form $\mathcal{O} = \{v \in V | f(v) \neq 0\}$ for some $f \in \mathcal{Q}(V)$ (e.g., $\mathcal{O} = K-\{0\} = \{\alpha \in K | \alpha \neq 0\}$) and let X denote the basic open set $\mathcal{O}$ with its structure ring $\mathcal{Q}(\mathcal{O})[1/f]$. (Cf. 1.5.31; X is complete by 1.5.32. In the specific example the structure ring is $K[X, 1/X]$ which can be identified concretely with the ring of finite Laurent series $\sum_{k=-m}^{n} a_k X^k$.) The identity map $X \to \mathcal{O}$ is clearly a weakening morphism and $\mathcal{O}$ is not complete by 1.5.12.

1.11.26. <u>Proposition.</u> Let $f : X \to Y$ be a morphism of ringed spaces. Then f factors uniquely in the form $f = j \circ W \circ \prod_{\mathcal{E}}$ where $\prod_{\mathcal{E}} : X \to X/\mathcal{E}$ is the projection of a ringed equivalence relation on X, $W : X/\mathcal{E} \to U$ is a weakening morphism, and $j : U \to Y$ is the inclusion of a ringed subspace U of Y into Y.

<u>Proof.</u> Clearly if such a factoring exists $U = f(X)$ and $\mathcal{E} = \mathcal{E}_f = \{(x_1, x_2) \in X \times X | f(x_1) = f(x_2)\} = \mathcal{E}_B$ where $B = f^*(\mathcal{Q}(Y)) \subseteq \mathcal{Q}(X)$ (cf. 1.11.12). On the other hand, this clearly gives such a factoring. ■

2.0. Conventions, Definitions, and Notations.

Henceforth we assume that the field K is infinite. In particular K may be any field of characteristic 0. If K has characteristic $p > 0$ we impose the additional restriction that K be perfect. We recall that a field K of characteristic $p > 0$ is called perfect if the Frobenius map $\alpha \mapsto \alpha^p$ (which always maps K isomorphically onto some subfield of K) is surjective, and hence an automorphism of K. Equivalently K is perfect if every element of K has a pth root in K.

An important consequence of K being infinite is that $K[X_1, \ldots, X_n]$ is strictly semi-simple (cf. 1.2.14). In fact we may now safely identify the algebra $K[X_1, \ldots, X_n]$ of formal polynomials in n-variables with the algebra $\mathcal{P}(K^n)$ of polynomial functions on K^n, by identifying X_i with the canonical projection of K^n on its ith factor and more generally identifying $f(X_1, \ldots, X_n)$, the formal polynomial, with the function $(\alpha_1, \ldots, \alpha_n) \mapsto f(\alpha_1, \ldots, \alpha_n)$. The reason we assume K perfect is as follows. Suppose $f(X_1, \ldots, X_n)$ is an element of $K[X_1, \ldots, X_n]$ such that the formal partial derivative of f with respect to each variable X_i is zero. Then the exponent of each X_i in each term of f is divisible by p, the characteristic of K. If we replace each coefficient of f by a pth root and divide all exponents of f by p we get a polynomial $g(X_1, \ldots, X_n)$ whose p^{th} power is $f(X_1, \ldots, X_n)$. In particular it follows that f is not irreducible.

We shall consider K^n as a ringed space over K with $\mathcal{P}(K^n)$ as structure ring. As such we refer to it as the n-<u>dimensional affine space over</u> K.

It is the prime example of an affine algebraic variety over K, a concept we define next.

2.0.1 <u>Definition</u>. A ringed space over K is called an (affine) <u>algebraic space</u> (over K) if its structure ring is finitely generated as a K-algebra. If in addition it is a complete ringed space then it is called an (affine) algebraic variety (over K). The Z-closed subsets of an <u>algebraic variety</u> are referred to as its algebraic subvarieties. The structures ring of an algebraic space S will usually be denoted by $\mathcal{P}(S)$, and elements of $\mathcal{P}(S)$ will be called <u>polynomial functions</u> on S. The set of ringed space morphisms from an algebraic space S_1 to an algebraic space S_2 will be denoted by $\mathcal{P}(S_1, S_2)$. Elements of $\mathcal{P}(S_1, S_2)$ will be called polynomial maps from S_1 to S_2.

2.0.1. <u>Remark</u>. The category of algebraic spaces and polynomial maps is a full subcategory of the category of ringed spaces. Thus all concepts meaningful for ringed spaces are meaningful when applied to algebraic spaces. For example an algebraic space is called irreducible if it is an irreducible space in its Z-topology or equivalently if and only if $\mathcal{P}(S)$ is an integral domain (cf. 1.7.18). We now remark on some basic properties of algebraic spaces which follow directly from the results of Part 1. In what follows we shall frequently refer to these properties without further justification.

In the first place algebraic spaces are Noetherian ringed spaces (1.8.4), and hence are Noetherian spaces with respect to their Z-topologies. In particular the Z-closed subsets of an algebraic space satisfy the descending chain condition, and more particularly the algebraic subvarieties of an algebraic variety satisfy the descending chain condition. It follows of course that algebraic spaces are compact in their Z-topologies - and in fact every subset is compact! If X_0 is an algebraic subspace of an algebraic space X then X_0 is an irreducible subspace of X if and only if $I(X_0) = \{f \in \mathcal{P}(X) | f \text{ vanishes on } X_0\}$ is a prime ideal of $\mathcal{P}(X)$, and X_0 is an irreducible component of X if and only if $I(X_0)$ is a minimal prime ideal of $\mathcal{P}(X)$. Moreover there are only finitely many minimal prime ideals of $\mathcal{P}(X)$ and each is of the form $I(X_0)$ for some unique irreducible component X_0 of X (cf. 1.8.6), i.e. the minimal prime ideals of X are strict radical ideals. In particular X is the finite union of its irreducible components $X_1, \ldots, X_n$. No X_i is included in the union of others, and $X = X_1 \cup \ldots \cup X_n$ is the unique way (up to order) to express X as the union of a finite number of Z-closed subspaces such that no one of them is included in any other (1.7.13).

2.0.2. Proposition. The n-dimensional affine space K^n is an irreducible algebraic variety.

Proof. Recall the elementary fact that if an algebra $\mathfrak{a}$ has no zero divisors (i.e., is an integral domain) then the same is true of the polynomial ring $\mathfrak{a}[X]$, so that by induction $K[X_1, \ldots, X_n]$ is an integral domain. Since $\mathcal{P}(K^n)$ is isomorphic to $K[X_1, \ldots, X_n]$ it follows that K^n is irreducible. For $i = 1, 2, \ldots, n$ let $x_i \in \mathcal{P}(K^n)$ be defined by $x_i(a_1, \ldots, a_n) = a_i$. Given $\varphi \in \mathcal{P}(K^n)^\wedge$ let $\varphi(x_i) = a_i$ and let $a = (a_1, \ldots, a_n)$. Since

$$Ev(a)(x_i) = x_i(a) = a_i = \varphi(x_i)$$

and $x_1, \ldots, x_n$ generate $\mathcal{P}(K^n)$ as a K-algebra, it follows that $Ev(a) = \varphi$, so $Ev : K^n \to \mathcal{P}(K^n)^\wedge$ is surjective and hence K^n is a complete ringed space and so an algebraic variety. ∎

2.0.3. Corollary. A ringed subspace S of K^n is an algebraic variety if and only if S is an algebraic subvariety of K^n. More generally a ringed subspace X_0 of any algebraic variety X is an algebraic variety if and only if X_0 is an algebraic subvariety of X.

Proof. Immediate from 1.5.12. ∎

2.0.4. Proposition. Let $(S, \mathcal{Q})$ be a ringed space over K and let $f : S \to K^m$ be a set map, say $s \mapsto (f_1(s), \ldots, f_m(s))$. Then f is a ringed space morphism $(S, \mathcal{Q}) \to (K^m, \mathcal{P}(K^m))$ if and only if $f_1, \ldots, f_m$ all belong to $\mathcal{Q}$.

Proof. Suppose $f_1, \ldots, f_m$ all belong to $\mathcal{Q}$ and let $g \in \mathcal{P}(K^m)$, say $g(a_1, \ldots, a_m) = G(a_1, \ldots, a_m)$ where $G(X_1, \ldots, X_m) \in K[X_1, \ldots, X_m]$. Then clearly $g \circ f = G(f_1, \ldots, f_m)$, which belongs to $\mathcal{Q}$ since $\mathcal{Q}$ is an algebra over K; so f is a ringed space morphism. Conversely if f is a ringed space morphism let $x_i \in \mathcal{P}(K^n)$ be defined by $x_i(a_1, \ldots, a_n) = a_i$. Then $x_i \circ f \in \mathcal{Q}$. But clearly $x_i \circ f = f_i$. ∎

2.0.5. Corollary. If $(S, \mathcal{Q})$ is a ringed space over K the $\mathcal{Q}$ is just the set of ringed space morphisms $f : (S, \mathcal{Q}) \to (K, \mathcal{P}(K))$. In particular if S is an algebraic space the $\mathcal{P}(S)$, the set of polynomial functions on S, is just the set $\mathcal{P}(S, K)$ of polynomial maps of S into K.

Proof. Take $m = 1$. ∎

2.0.6. Corollary. K^m is the product of m copies of K in the category of ringed spaces over K.

2.0.7. Corollary. The set $\mathcal{P}(K^n, K^m)$ of polynomial maps $K^n \to K^m$ is the set of maps of the form $(a_1, \ldots, a_n) \mapsto (f_1(a_1, \ldots, a_n), \ldots, f_m(a_1, \ldots, a_n))$ where $f_1, \ldots, f_m \in K[X_1, \ldots, X_n]$.

2.0.8. Example. By definition $\mathcal{P}(K^n)$ is generated by the n-functions $X_1, \ldots, X_n$ (where $X_i(a_1, \ldots, a_n) = a_i$). Since these n-functions are a basis for the linear functionals on K^n, we can equally well say that $\mathcal{P}(K^n)$ is the algebra of K-valued functions on K^n generated by the dual space $(K^n)^*$ of K^n. If V is any finite dimensional vector space over K we define $\mathcal{P}(V)$, the algebra of polynomial functions on V, to be the algebra of K-valued functions on V generated by the dual space V^* of V. Any basis for V^* then clearly generates $\mathcal{P}(V)$, so V becomes an algebraic space over K. Let V_1 and V_2 be finite dimensional vector spaces over K and let $T : V_1 \to V_2$ be a linear map. The homomorphism $K^T : K^{V_2} \to K^{V_1}$, namely $g \mapsto g \circ T$, restricts to the adjoint map $T^* : V_2^* \to V_1^*$. Since V_i^* generates $\mathcal{P}(V_i)$ it follows that K^T maps $\mathcal{P}(V_2)$ into $\mathcal{P}(V_1)$, i.e. that T is a ringed space morphism $V_1 \to V_2$. If T is a vector space isomorphism it follows that T is also a ringed space isomorphism. In particular if $\dim(V) = n$ there is a vector space isomorphism $T : V \to K^n$ and it follows that V is isomorphis to K^n as a ringed space. In particular V is an irreducible algebraic variety.

2.0.9. Remark. So far we have been careful to distinguish between the formal polynomial $P(X_1, \ldots, X_n)$ and the polynomial function $K^n \to K$ given by $(a_1, \ldots, a_n) \mapsto P(a_1, \ldots, a_n)$. To avoid unnecessary locutions we shall henceforth regard this isomorphism of $K[X_1, \ldots, X_n]$ with $\mathcal{P}(K^n)$ as an identification. In particular we regard X_i both as a formal polynomial and the projection $(a_1, \ldots, a_n) \mapsto a_i$.

As a corollary to the above identification we will identify $K[X_1, \ldots, X_n]^\wedge$ with K^n; that is we identify $\varphi \in K[X_1, \ldots, X_n]^\wedge$ with $(\varphi(X_1), \ldots, \varphi(X_n)) \in K^n$. We recall that the inverse map is $Ev : K^n \to K[X_1, \ldots, X_n]^\wedge$ where $Ev(a)(P) = P(a_1, \ldots, a_n)$. This latter identification in turn entails more or less automatically an identification of the Z-closed subsets of $K[X_1, \ldots, X_n]^\wedge$ with the algebraic subvarieties of K^n. Specifically, given an ideal $\mathcal{J}$ of $K[X_1, \ldots, X_n]$ the corresponding Z-closed subset $V(\mathcal{J})$ of $K[X_1, \ldots, X_n]^\wedge$ is the algebraic subvariety of K^n defined by

$$V(\mathcal{J}) = \{(a_1, \ldots, a_n) \in K^n \mid P(a_1, \ldots, a_n) = 0 \text{ for all } P \in \mathcal{J}\}.$$

And given a subset S of K^n we denote by $I(S)$ the corresponding strict radical ideal in $K[X_1, \ldots, X_n]$,

$$I(S) = \{P(X_1, \ldots, X_n) \in K[X_1, \ldots, X_n] \mid P \text{ vanishes on } S\}.$$

2.0.10. Remark. In 1.5.22 we saw that products existed in the category of all ringed spaces over K. Given $(S_i, \mathfrak{A}(S_i))$ $(i=1,2)$ their product was defined as $(S_1 \times S_2, \mathcal{Q}(S_1) \otimes \mathcal{Q}(S_2))$, where $f \otimes g$ is the function $(s_1, s_2) \mapsto f(s_1)g(s_2)$. Now if S_1 and S_2 are algebraic spaces we have generators say $f_1, \ldots, f_n$ for $\mathcal{Q}(S_1)$ and $g_1, \ldots, g_m$ for S_2 and then $\{f_i \otimes g_i\}$ generate $\mathcal{Q}(S_1) \otimes \mathcal{Q}(S_2)$ so $S_1 \times S_2$ is also an algebraic space. Hence products exist in the category of algebraic spaces over K. We note that by 1.7.22 the product of algebraic spaces is irreducible if and only each factor is irreducible.

In the same way, and even more trivially we see from 1.5.25 that sums exist in the category of algebraic spaces. Moreover by 1.5.24 and 1.5.25, products and sums also exist in the subcategory of algebraic varieties.

2.0.11. Definition. Let G be an algebraic space over K whose underlying set is a group. We call G an algebraic group if the morphism $(x, y) \mapsto xy^{-1}$ of $G \times G \to G$ is a polynomial map, i.e. if given $f \in \mathcal{P}(G)$ there exist $g_1, \ldots, g_n, h_1, \ldots, h_n$ in $\mathcal{P}(G)$ such that

$$f(xy^{-1}) = \sum_{i=1}^{n} g_i(x)h_i(y)$$

for all $x, y \in G$. If X is an algebraic space over K whose underlying set is a G-set (cf. 1.5.26) then we call X an algebraic G-space if the action of G on X is a polynomial map $G \times X \to X$.

2.0.12. Remark. Note that there is really nothing new in the definition 2.0.11. An algebraic group is just a ringed group (cf. 1.5.26) which happens to be an algebraic space. Similarly an algebraic G-space is just a ringed G-space for an algebraic group G which happens to be an algebraic space.

2.1. Generating Points.

2.1.1. Definition. Let S be an algebraic space over K and let $\xi = (\xi_1, \ldots, \xi_n)$ be an ordered n-tuple of elements of $\mathcal{P}(S)$. Define $h^\xi : K[X_1, \ldots, X_n] \to \mathcal{P}(S)$ by $P(X_1, \ldots, X_n) \mapsto P(\xi_1, \ldots, \xi_n)$. We call ξ a generating point for S if h^ξ is surjective, i.e. if $\xi_1, \ldots, \xi_n$ generate $\mathcal{P}(S)$ as an algebra over K. In this case we denote $\ker(h^\xi)$ by $\mathcal{J}^\xi$, so we have an exact sequence

$$0 \to \mathcal{J}^\xi \to K[X_1, \ldots, X_n] \to \mathcal{P}(S) \to 0,$$

and we denote by $E^\xi : S \to K^n$ the map $x \mapsto (\xi_1(x), \ldots, \xi_n(x))$. The image of E^ξ in K^n is denoted by S^ξ. We consider S^ξ as a ringed subspace of K^n.

2.1.2. Proposition. Let S be an algebraic space over K. If $\xi = (\xi_1, \ldots, \xi_n)$ is a generating point for S then $E^\xi : S \to K^n$ is an injective polynomial map. In fact $E^\xi_* : \mathcal{P}(K^n) \to \mathcal{P}(S)$, namely $P \mapsto P \circ E^\xi$, is just $h^\xi : K[X_1, \ldots, X_n] \to \mathcal{P}(S)$ under the canonical identification of $\mathcal{P}(K^n)$ with $K[X_1, \ldots, X_n]$. Equivalently, E^ξ is the composition of $Ev : S \to \mathcal{P}(S)^\wedge$, $(h^\xi)^\wedge : \mathcal{P}(S)^\wedge \to K[X_1, \ldots, X_n]^\wedge$, and the canonical identification of $K[X_1, \ldots, X_n]^\wedge$ with K^n (namely, $\varphi \mapsto (\varphi(X_1), \ldots, \varphi(X_n))$).

Proof. Since $\xi_i \in \mathcal{P}(S)$, that E^ξ is a polynomial map is immediate from 2.0.4. In fact given $P \in K[X_1, \ldots, X_n]$ $E^\xi_*(P) = P \circ E^\xi = P(\xi_1, \ldots, \xi_n) = h^\xi(P)$ is obvious. Since h^ξ is surjective, E^ξ is injective. ∎

2.1.3. Theorem. Let S be an algebraic space over K and let $\xi = (\xi_1, \ldots, \xi_n)$ be a generating point for S. Then:

(1) $\mathcal{J}^{\xi} = I(S^{\xi}) = \{P \in K[X_1, \ldots, X_n] \mid P$ vanishes on $S^{\xi}\}$.

(2) $V(\mathcal{J}^{\xi})$ is the Z-closure of S^{ξ} in K^n.

(3) $E^{\xi} : S \to K^n$ is an injective polynomial map, and is a (ringed space) isomorphism with its image S^{ξ} (considered as a ringed subspace of K^n).

(4) S is an algebraic variety if and only if S^{ξ} is an algebraic subvariety of K^n.

Proof. $P \in \mathcal{J}^{\xi} \Leftrightarrow P(\xi_1, \ldots, \xi_n) = 0$

$\Leftrightarrow P(\xi_1, \ldots, \xi_n)(p) = 0$ all $p \in S$

$\Leftrightarrow P(\xi_1(p), \ldots, \xi_n(p)) = 0$ all $p \in S$

$\Leftrightarrow P$ vanishes on S^{ξ}.

This proves (1), and since the Z-closure of S^{ξ} is $V(I(S^{\xi}))$ it also proves (2). That E^{ξ} is an injective polynomial map is part of 2.1.2. Let $\widetilde{E}^{\xi}$ denote E^{ξ} considered as a map into its image S^{ξ}, the latter regarded as a ringed subspace of K^n, and let $i : S^{\xi} \to K^n$ be the inclusion map, so that $E^{\xi} = i \circ \widetilde{E}^{\xi}$. Note that $i_* : \mathcal{P}(K^n) \to \mathcal{P}(S^{\xi})$, namely $P \mapsto P \circ i$, is really just the restriction map $P \mapsto P|S^{\xi}$, so that $\ker(i_*) = I(S^{\xi})$. Now by 2.1.2, $h^{\xi} = E^{\xi}_*$, so $h^{\xi} = \widetilde{E}^{\xi}_* \circ i_*$, i.e. we have a commutative diagram:

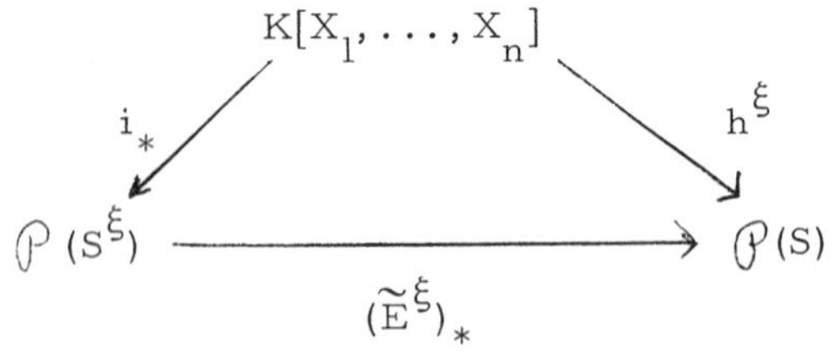

Since by (1) we have:

$$\ker(i_*) = I(S^\xi) = \mathcal{J}^\xi = \ker(h^\xi)$$

it follows that $(\widetilde{E}^\xi)_*$ is an isomorphism, which proves (3). It follows from (3) that S^ξ is an algebraic variety (i.e. complete as a ringed space) if and only if S is, so that (4) follows from 2.0.3. ■

2.1.4. Corollary. Any affine algebraic space S over K is isomorphic to a ringed subspace X of some n-dimensional affine space K^n. Moreover S is an affine algebraic variety over K if and only if X is an algebraic subvariety of K^n.

2.1.5. Lemma. If X is a ringed space over K and X_0 is a ringed subspace of X, then a set map $X_0 \to K^n$ is a ringed space morphism if and only if it extends (not necessarily uniquely) to a ringed space morphism $X \to K^n$.

Proof. Immediate from 2.0.4 and the fact that (by definition) $f : X_0 \to K$ is in the structure ring $\mathcal{A}(X_0)$ of X_0 if and only if it is the restriction to X_0 of some element of $\mathcal{A}(X)$. ■

2.1.6. Theorem. Let S_1 and S_2 be algebraic spaces over K and let $\xi = (\xi_1, \ldots, \xi_m)$ be a generating point for S_1 and $\eta = (\eta_1, \ldots, \eta_n)$ a generating point for S_2. If $f : S_1 \to S_2$ is a set map then the following are equivalent:

1) f is a ringed space morphism, i.e. f belongs to the set $\mathcal{P}(S_1, S_2)$ of polynomial maps.

2) There exist $F_1, \ldots, F_n$ in $K[X_1, \ldots, X_m]$ such that the polynomial

map $F : K^m \to K^n$ defined by $(a_1, \ldots, a_m) \mapsto (F_1(a_1, \ldots, a_m), \ldots, F_n(a_1, \ldots, a_m))$ makes the following diagram commute.

$$\begin{array}{ccc} S_1 & \xrightarrow{\quad f \quad} & S_2 \\ {\scriptstyle E^\xi}\big\downarrow & & \big\downarrow{\scriptstyle E^\eta} \\ K^m & \xrightarrow[\quad F \quad]{} & K^n \end{array}$$

Proof. By 2.1.3 there is a uniquely determined map $\tilde{f} : S_1^\xi \to S_2^\eta$ such that $E^\eta \circ f = \tilde{f} \circ E^\xi$ and f is a ringed space morphism if and only if $\tilde{f}$ is. If $i : S_2^\eta \to K^n$ is the inclusion map then, since S_2^η is a ringed subspace of K^n, $\tilde{f} : S_1^\xi \to S_2^\eta$ is a ringed space morphism if and only if $i \circ \tilde{f} : S_1^\xi \to K^n$ is a ringed space morphism. By 2.1.5 the latter is the case if and only if there is a ringed space morphism $F : K^m \to K^n$ which extends $i \circ \tilde{f}$, and by 2.0.7 this is equivalent to (2). ∎

2.1.7. Remark. With the notation of the preceding theorem, if $f : S_1 \to S_2$ is a polynomial map then one gets the polynomials $F_i(X_1, \ldots, X_m) \in K[X_1, \ldots, X_n]$ as follows. Since $g \circ f \in \mathcal{P}(S_1)$ for all $g \in \mathcal{P}(S_2)$, in particular $\eta_i \circ f \in \mathcal{P}(S_1)$. Since $\xi_1, \ldots, \xi_m$ generate $\mathcal{P}(S_1)$ there is a polynomial $F_i(X_1, \ldots, X_m)$ such that $\eta_i \circ f = F_i(\xi_1, \ldots, \xi_m)$. Then note that $\eta_i(f(p)) = F_i(\xi_1(p), \ldots, \xi_m(p))$ for all $p \in S_1$, which says $E^\eta \circ f = F \circ E^\xi$.

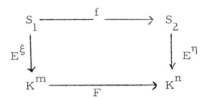

2.2. Review of Some Commutative Algebra.

In this section we review some of the concepts of commutative algebra which play an important role in algebraic geometry. Our aim will be mainly to develop notation and collect in one place the results we shall need later. In particular, while we shall attempt to cover the facts we need in a logical order, we shall not always give complete proofs. For details that are omitted the reader is referred to one of the standard algebra texts, for example [17], [33], or [35].

2.2.1. Notation. In this section A will denote an integral domain (with unit), B a subring of A (with unit) and F and K the respective fields of quotients of A and B (so that F is an extension field of K). Note in particular we may have $A = F$ and/or $B = K$. As usual $B[X]$ denotes the ring of polynomials in X with coefficients in B. If $\alpha \in A$ then $B[\alpha]$ denotes the subring of A generated by B and α (i.e. the set of all $P(\alpha)$, for $P(X)$ in $B[X]$). And if $\alpha \in F$ then $K(\alpha)$ denotes the smallest extension field of K containing α (i.e. the set of $P(\alpha)/Q(\alpha)$ where $P(X)$ and $Q(X)$ are in $K[X]$ and $Q(\alpha) \neq 0$). If $\alpha \in A$ then $\mathscr{I}_B(\alpha)$ will denote the ideal of $P(X)$ in $B[X]$ such that $P(\alpha) = 0$, so that we have an exact sequence

$$0 \to \mathscr{I}_B(\alpha) \to B[X] \to B[\alpha] \to 0 .$$

The element α of A is called transcendental over B if the map $P(X) \to P(\alpha)$ of $B[X]$ onto $B[\alpha]$ is an isomorphism, or equivalently if $\mathscr{I}_B(\alpha) = (0)$. In the contrary case α is called algebraic over B. Thus α is algebraic over

B if and only if there is a nonzero polynomial $P(X)$ in $B[X]$ such that $P(\alpha) = 0$; if in addition there exists such a $P(X)$ which is monic (i.e. with leading coefficient 1) then α is called integral over B. The set of $\alpha \in A$ which are integral over B will be denoted by $Int(A/B)$ and called the integral closure of B in A, and B will be called integrally closed in A if $Int(A/B) = B$. Similarly the set of $\alpha \in A$ which are algebraic over B will be denoted by $Alg(A/B)$ and called the algebraic closure of B in A, and B will be called algebraically closed in A if $Alg(A/B) = B$.

2.2.2. Remark. In case $B = K$ (i.e. B is a field) the distinction between being integral over B and merely being algebraic over B evaporates; for if $P(X)$ is any non-zero element of $\mathcal{J}_B(\alpha)$, then dividing each coefficient of $P(X)$ by the leading coefficient gives a monic element of $\mathcal{J}_B(\alpha)$. In general however the distinction is real and important. For example every element x of K is clearly algebraic over B (if $x = b_0/b_1$ with $b_0, b_1 \in B$ then x satisfies the polynomial equation $P(x) = 0$ where $P(X) \in B[X]$ is $b_1X - b_0$), that is $Alg(K/B) = K$. On the other hand if $B = \mathbb{Z}$, or a polynomial ring over a field, or more generally a unique factorization domain (UFD) then $Int(K/B) = B$, i.e. a UFD is always integrally closed in its field of quotients [the proof in essence goes back to the Pythagnean observation that $X^2 = 2$ has no solution in the rationals. Suppose $x = b_0/b_1$ satisfies the monic equation $X^n = \beta_0 + \beta_1 X + \ldots + \beta_{n-1}X^{n-1}$. Since B is a UFD we can suppose that the representation of x as b_0/b_1 is in "lower terms", i.e. no irreducible of B divides both b_0 and b_1. It will then suffice to prove that b_1 is a unit of B so that $x \in B$. Substituting x in the equation and multiplying by b_1^n, we see that if ρ is any irreducible factor of b_1 then ρ divides b_0^n and hence ρ divides b_0, contradicting the "lowest terms" assumption. Thus b_1 has no irreducible factors and is a unit].

2.2.3. Proposition. An element of an integral domain is algebraic over a subring if and only if it is algebraic over the field of quotients of that subring. That is $\mathrm{Alg}(A/B) = A \cap \mathrm{Alg}(F/K)$.

Proof. It is trivial that if $\alpha \in A$ is algebraic over B it is algebraic over K. Conversely if $P(\alpha) = 0$ where $0 \neq P(X) = c_0 + c_1X + \ldots + c_nX^n \in K[X]$, then each c_j can be written as a quotient of elements of B, say $c_j = n_j/d_j$ where of course $d_j \neq 0$ and hence $d = d_0d_1\ldots d_n \neq 0$. Then "clearing of denominators", i.e. multiplying $P(X)$ by d, gives a non-zero polynomial in $B[X]$ having α as a root. ■

2.2.4. Proposition. For $\alpha \in A$ the following are equivalent:

1) α is integral over B, i.e. $\alpha \in \mathrm{Int}(A/B)$.

2) $B[\alpha]$ is finitely generated as a B-module.

3) α is contained in a subring R of A which is a finitely generated B-module.

Proof. ((1) ⇒ (2)). By assumption α satisfies an equation $\alpha^n = b_0 + b_1\alpha + \ldots + b_{n-1}\alpha^{n-1}$ where the b_i are in B. Then $\alpha^{n+1} = b_0\alpha + \ldots + b_n\alpha^n$, and substituting for α^n in the latter equation from the former shows that not only α^n but also α^{n+1} is contained in the finitely generated B module $B_1 + B\alpha + \ldots + B\alpha^{n-1}$. Inductively it follows in the same way that each α^{n+k} is contained in this B-module, hence so is $B[\alpha]$.

((2) ⟹ (3)) is trivial.

((3) ⟹ (1)). Let $R = Bu_1 + \dots + Bu_n$.

Since R is a ring $\alpha R \subseteq R$ and so $\alpha u_i \in R$, hence $\alpha u_i = \sum_{j=1}^{n} b_{ij}u_j$ for some $b_{ij} \in B$, or equivalently $\sum_{j=1}^{n} (b_{ij} - \alpha\delta_{ij})u_j = 0$ for $i = 1, 2, \dots, n$. Now if the u_i are all zero then $R = (0)$ so $\alpha = 0$ and we are done. Otherwise $\det(b_{ij} - \alpha\delta_{ij}) = 0$ and $\det(b_{ij} - X\delta_{ij})$ is a monic polynomial in $B[X]$ having α as a root. ∎

2.2.5. Proposition. Let R be the subring of A generated by two subrings R_1 and R_2. If $u_1, \dots, u_m$ generate R_1 as a B module and $v_1, \dots, v_n$ generate R_2 as a B module then the u's, v's and products $u_k v_j$ generate R as a B-module.

Proof. Trivial. ∎

2.2.6. Proposition. Int(A/B), the integral closure of B in A, is a subring of A. Moreover it is integrally closed in A.

Proof. It is immediate from 2.2.4 and 2.2.5 that Int(A/B) is a ring. Let α in A be integral over Int(A/B) = $\overline{B}$. Then α satisfies $\alpha^n + a_{n-1}\alpha^{n-1} + \dots + a_0$ where $a_0, a_1, \dots, a_{n-1}$ are in $\overline{B}$. Now each of the a_i belong to a subring R_i of A which is finitely generated as a B module. By an easy induction from 2.2.5, the ring R generated by the R_i, which of course contains $a_0, a_1, \dots, a_{n-1}$, is also finitely generated as a B-module. Say $R = Bu_1 + \dots + Bu_m$. Clearly α is integral over R, so there is a subring Q of A containing α such that Q is finitely generated as an R

module; say $Q = Rv_1 + \ldots + Rv_k$. Then as a B module Q is the sum of the cyclic submodules Bu_iv_j, so every element of the ring Q, and in particular α, is integral over B. ∎

2.2.7. Proposition. An integral domain D which includes a field K and is finite dimensional over K is a field.

Proof. Given $d \neq 0$ in D, multiplication by d is a K-linear map of D to itself. Moreover $dx = 0$ implies $x = 0$ so the map is non-singular, and since D is finite dimensional over K it is surjective. In particular $dx = 1$ has a solution x in D. ∎

2.2.8. Corollary. If α is an element of F algebraic over K then $K[\alpha]$ is a field and hence $K[\alpha] = K(\alpha)$. Conversely if $K[\alpha]$ is a field then α is algebraic over K.

Proof. The first conclusion is immediate from 2.2.4 and 2.2.7. If α is transcendental over K then $K[\alpha]$ is isomorphic to the ring $K[X]$ of formal polynomials in X. If $P(X) \in K[X]$ and degree $(P) \geq 1$ then for any $Q(X) \neq 0$ in $K[X]$, degree (PQ) is equal to degree (P) + degree (Q) which is greater than 0, hence $P(X)Q(X) \neq 1$ and $P(X)$ does not have an inverse in $K[X]$. ∎

2.2.9. Corollary. Alg (F/K), the algebraic closure of K in F, is a subfield of F and is algebraically closed in F.

Proof. Recall from 2.2.2 that Alg(F/K) is the same as Int(F/K). The result then follows from 2.2.6, 2.2.4, and 2.2.8. ∎

2.2.10. Proposition (Nullstellensatz Lemma). If $a_1, \ldots, a_n$ are elements of F then the K algebra $K[a_1, \ldots, a_n]$ they generate is a field if and only the a_i are all algebraic over K.

Proof. The case n = 1 is just 2.2.8 so we proved by induction. If $a_1, \ldots, a_n$ are algebraic over K then a fortiori a_n is algebraic over the field $K[a_1, \ldots, a_{n-1}]$ so $K[a_1, \ldots, a_n] = K[a_1, \ldots, a_{n-1}][a_n]$ is a field. Now assume $K[a_1, \ldots, a_n]$ is a field so that in particular it includes $K(a_1)$ and is equal to $K(a_1)[a_2, \ldots, a_n]$. By induction $a_2, \ldots, a_n$ are algebraic over $K(a_1)$ and the inductive step will follow from 2.2.9 if we can prove a_1 is algebraic over K. Now a_2 is algebraic over $K[a_1]$ by 2.2.3 so we have $a_n(a_1)a_2^n + \ldots + a_0(a_1) = 0$, where $a_n(X)$ is a polynomial such that $a_n(a_1) \neq 0$. Multiplying this equation by $a_n(a_1)^{n-1}$ we see that $p_2(X) = a_n(X)$ is a polynomial such that $p_2(a_1) \neq 0$ and $p_2(a_1)a_2$ is integral over $K[a_1]$. Since a_1 is certainly integral over $K[a_1]$ and the elements integral over $K[a_1]$ from a ring it follows that any product of $p_2(X)$ with a polynomial $Q(X)$ such that $Q(a_1) \neq 0$ has the same property. Similarly we can find a polynomial $p_j(X)$, $j = 3, \ldots n$ such that $p_j(a_1) \neq 0$ and $p_j(a_1)a_j$ is integral over $K[a_1]$. Then $A(X) = p_2(X) \ldots p_n(X)$ has the property that $A(a_1) \neq 0$ and $A(a_1)a_j$ is integral over $K[a_1]$ for $j = 2, \ldots, n$. Let $C(X)$ be any irreducible polynomial that does not divide $A(X)$. Suppose now that a_1 were transcendental over K. In particular $C(a_1) \neq 0$ so $C(a_1)^{-1} \in K(a_1) \subseteq K[a_1, \ldots, a_n]$. If $C(a_1)^{-1} = P(a_1, \ldots, a_n)$ and d is the total degree of P then clearly $C(a_1)^{-1}A(a_1)^d$ is a polynomial in a_1, $A(a_1)a_2, \ldots, A(a_1)a_n$ so it is integral over $K[a_1]$. But since a_1 is transcendental, $K[a_1]$ is isomorphic to the polynomial ring $K[X]$ and hence (cf. 2.2.2) it is integrally closed in its field of quotients, i.e. $C(a_1)^{-1}A(a_1)^d = G(a_1)$ where $G(X) \in K[X]$. Then $A(a_1)^d = C(a_1)G(a_1)$ and since a_1 is transcendental, $A(X)^d = C(X)G(X)$. This $C(X)$ divides $A(X)^d$ and since $C(X)$ is irreducible $C(X)$ divides $A(X)$, a contradiction.

■

2.2.11. Remark. We recall the basic properties of the ring $K[X]$ of polynomials in one variable over a field K. It is a UFD, and in fact every element of $K[X]$ factors uniquely into an element of K (a unit of the ring) and a product of monic irreducible polynomials (unique except for their order). It is a principal ideal domain. Every ideal $\mathcal{J}$ in $K[X]$ has a unique monic generator $M(X)$. $M(X)$ is characterized as the monic polynomial of least degree in $\mathcal{J}$. If d is the degree of $M(X)$ then clearly $1, X, \ldots, X^{d-1}$ are linearly independent over K modulo $\mathcal{J}$. Since by the Euclidean algorithm every element of $K[X]$ is equivalent modulo $M(X)$ to a polynomial of degree less than d it follows that the residue classes of $1, X, \ldots, X^{d-1}$ form a K basis for $K[X]/\mathcal{J}$, so in particular the latter has dimension d over K. It is trivial that $K[X]/\mathcal{J}$ is an integral domain if and only if $M(X)$ is irreducible, and it then follows from 2.2.7 that in this case $K[X]/\mathcal{J}$ is in fact a field.

If $a \in F$ is algebraic over K we denote by $M_a(X)$ the monic generator of $\mathcal{J}_K(a)$. $M_a(X)$ is called the minimal polynomial of a and its degree is called the degree of α over K, denoted by $\text{degree}_K(\alpha)$. Of course $M_\alpha(X)$ is irreducible and $K(\alpha) = K[\alpha] = K[X]/(M_\alpha(X))$. Thus $[K(\alpha) : K]$, the dimension of $K(\alpha)$ over K, is just $\text{degree}_K(\alpha)$.

2.2.12. Remark. Now suppose B is a UFD (unique factorization domain). Given $P(X) \in K[X]$ (where K as usual is the field of fractions of B) we can write the coefficients of P(X) over a common denominator and then factor the GCD out of the resulting numerators. Thereby we get a factoring $P(X) = C_P P_0(X)$ where $P_0(X) \in B[X]$, the coefficients of $P_0(X)$ are relatively prime in B, and C_P is a non-zero element of K (we assume $P \neq 0$). C_P is called a content for P(X) and it is clear that C_P is determined up to a unit of B. According to Gauss' lemma [, Chapt. V, §6] if Q(X) is another non-zero element of K[X] and C_Q is a content for Q(X) then $C_P C_Q$ is a content for P(X)Q(X). Note also that if $P(X) \in B[X]$ then C_P is just a GCD of the coefficients of P(X) and in particular it is an element of B. If P(X) is an irreducible (prime) element of B[X] the clearly the coefficients of P(X) are relatively prime. Moreover P(X) is also irreducible in K[X]! For suppose P(X) were the product of two polynomial Q(X) and R(X) of positive degree in K[X]. Now $P(X) = (C_Q C_R) Q_0(X) R_0(X)$ where Q_0 and R_0 are elements of B[X] having the same degrees as Q and R respectively. Moreover by Gauss' lemma $(C_Q C_R)$ differs from $C_P = 1$ by a unit of B, i.e. $(C_Q C_R) \in B$, so P(X) factors non-trivially over B[X], a contradiction. Conversely it is trivial that if $P(X) \in B[X]$ is irreducible as an element of $K[X]$ and its coefficients are relatively prime in B then $P(X)$ has no non-trivial factorizations in $B[X]$. Since we know $K[X]$ is a UFD it follows easily that $B[X]$ is also a UFD. By induction it follows that $B[X_1, \ldots, X_n]$ is a UFD. In particular if K is any field then $K[X_1, \ldots, X_n]$ is a UFD.

2.2.13. Proposition. Assume B is a unique factorization domain and let $\alpha \in A$ be algebraic over B. Then $\mathcal{I}_B(\alpha) = \{P(X) \in B[X] | P(\alpha) = 0\}$ is a principal prime ideal in $B[X]$. Moreover the generator of $\mathcal{I}_B(\alpha)$ (which is clearly unique to within a unit of B) is irreducible in $K[X]$ and its coefficients are relatively prime in B.

Proof. Let $M(X)$ be the minimal polynomial of α over K, so that $M(X)$ is irreducible in $K[X]$. Then if we write $M(X) = C_M M_0(X)$ where C_M is a content for M, then M_0 has relatively prime coefficients and is irreducible in $B[X]$ and in $K[X]$ (cf. 2.2.12). Thus it will suffice to show that $M_0(X)$ generates $\mathcal{I}_B(\alpha)$. Now if $Q(X) \in \mathcal{I}_B(\alpha)$ then $Q(\alpha) = 0$ so $Q(X) \in \mathcal{I}_K(\alpha) = (M(X))$ i.e. $Q(X) = M(X)R(X)$, where $R(X) \in K[X]$. Then $Q(X) = C_M C_R M_0(X) R_0(X)$ where C_R is a content for R, $R_0(X) \in B[X]$ and by Gauss' lemma $C_M C_R$ is a content for $Q(X)$. Since $Q(X) \in B[X]$ it follows that $C_M C_R \in B$ so $C_M C_R R_0(X) \in B[X]$, i.e. $M_0(X)$ divides $Q(X)$ in $B[X]$. ∎

2.2.14. Definition. A finite set of elements $\alpha_1, \ldots, \alpha_n$ of F is called algebraically independent over K if the map $P(X_1, \ldots, X_n) \mapsto P(\alpha_1, \ldots, \alpha_n)$ of $K[X_1, \ldots, X_n]$ into F is a monomorphism, i.e. if the only $P(X_1, \ldots, X_n)$ such that $P(\alpha_1, \ldots, \alpha_n) = 0$ is the zero polynomial. In the contrary case

$\alpha_1, \ldots, \alpha_n$ are called <u>algebraically dependent</u> over K. An element α of F is said to <u>depend algebraically on</u> $\alpha_1, \ldots, \alpha_n$ (over K) if α is algebraic over $K[\alpha_1, \ldots, \alpha_n]$ (or what is the same by 2.2.3, if α is algebraic over $K(\alpha_1, \ldots, \alpha_n)$). (Note that the latter is <u>stronger</u> than just requiring that $\alpha_1, \ldots, \alpha_n, \alpha$ be algebraically dependent over K. For example if $\alpha_1, \ldots, \alpha_n$ are in K and α is transcendental over K then $\alpha_1, \ldots, \alpha_n, \alpha$ is algebraically dependent over K but α does <u>not</u> depend algebraically on $\alpha_1, \ldots, \alpha_n$ over K).

2.2.15. <u>Proposition</u>. If each of $\beta_1, \ldots, \beta_m$ depends algebraically on $\alpha_1, \ldots, \alpha_n$ then any element of F which depends algebraically on $\beta_1, \ldots, \beta_m$ also depends algebraically on $\alpha_1, \ldots, \alpha_n$.

<u>Proof</u>. By 2.2.9 the algebraic closure of $K[\alpha_1, \ldots, \alpha_n]$ in F is algebraically closed in F. ∎

2.2.16. <u>Proposition</u>. If α depends algebraically on $\alpha_1, \ldots, \alpha_n$ but not on $\alpha_1, \ldots, \alpha_{n-1}$ then α_n depends algebraically on $\alpha_1, \ldots, \alpha_{n-1}, \alpha$.

<u>Proof</u>. By assumption $\sum_{i=0}^{n} C_i \alpha^i = 0$ where $C_0, \ldots, C_n$ are elements of $K[\alpha_1, \ldots, \alpha_n]$, not all zero. Let $c_i = C_i(\alpha_1, \ldots, \alpha_n)$ where $C_i(X_1, \ldots, X_n) \in K[X_1, \ldots, X_n]$. Then the C_i are not all zero and hence $P(X_1, \ldots, X_n, Z) = \sum_{i=0}^{n} C_i(X_1, \ldots, X_n)Z^n$ is not the zero element of $K[X_1, \ldots, X_n, Z]$. Writing $P(X_1, \ldots, X_n, Z) = \sum_{i=0}^{d} Q_i(X_1, \ldots, X_{n-1}, Z)X_n^i$ not all the Q_i are zero, and since α does not depend algebraically on $\alpha_1, \ldots, \alpha_{n-1}$ it follows that not all of the $Q_i(\alpha_1, \ldots, \alpha_{n-1}, \alpha)$ are zero. Since $0 = \sum_{i=0}^{d} Q_i(\alpha_1, \ldots, \alpha_{n-1}, \alpha)\alpha_n^i$ it follows that α_n is algebraic over $K[\alpha_1, \ldots, \alpha_{n-1}, \alpha]$. ∎

2.2.17. Proposition (Exchange Theorem). Let $\alpha_1, \ldots, \alpha_n$ be elements of A algebraically independent over K. Let $\beta_1, \ldots, \beta_m$ be elements of A such that every element of A depends algebraically on $\beta_1, \ldots, \beta_m$. Then $n \leq m$ and in fact every element of A depends algebraically on $\alpha_1, \ldots, \alpha_n$ and some $(m-n)$-element subset of $\{\beta_1, \ldots, \beta_m\}$.

Proof. For $n = 0$ this is trivial, so we proceed by induction assuming the result when n is replaced by n-1. Then renumbering the β's it follows that every element of A depends algebraically on the set

$$S = \{\alpha_1, \ldots, \alpha_{n-1}, \beta_1, \ldots, \beta_{m-n+1}\} .$$

We must prove that there is an index j with $1 \leq j \leq m - n + 1$ such that every element of A depends algebraically on $S \cup \{\alpha_n\} - \{\beta_j\}$. By 2.2.15 it will suffice to show that every element of S depends algebraically on $S \cup \{\alpha_n\} - \{\beta_j\}$. It will even suffice to get a β_j depending algebraically on $S \cup \{\alpha_n\} - \{\beta_j\}$, since the other elements of S are contained in $S \cup \{\alpha_n\} - \{\beta_j\}$, so clearly algebraically dependent on it. Since $\alpha_n \in A$ it depends algebraically on S, so $S \cup \{\alpha_n\} = \{\alpha_1, \ldots, \alpha_n, \beta_1, \ldots, \beta_{m-n+1}\}$ is algebraically dependent. Since $\alpha_1, \ldots, \alpha_n$ are algebraically independent it follows that $m - n + 1 \geq 1$, or $n \leq m$, and it also follows from 2.2.16 then if j is the least index such that $\alpha_1, \ldots, \alpha_n, \beta_1, \ldots, \beta_j$ is algebraically dependent then (even if $j=1$) β_j depends algebraically on $\alpha_1, \ldots, \alpha_n, \beta_1, \ldots, \beta_{j-1}$. A fortiori β_j depends algebraically on the larger set $S \cup \{\alpha_n\} - \{\beta_j\}$. ∎

2.2.18. Definition. If for each positive integer n there exist $\alpha_1, \ldots, \alpha_n$ in A which are algebraically independent over K then we define the transcendence degree or A over K to be ∞. Otherwise the transcendence degree of A over K is the largest integer n for which such an algebraically independent set exists. Any algebraically independent set with this maximum number of elements is called a transcendence basis for A over K.

2.2.19. Proposition. A necessary and sufficient condition for A to have finite transcendence degree over K is that there exist a subset $\alpha_1, \ldots, \alpha_n$ of A such that every element of A depends algebraically on $\alpha_1, \ldots, \alpha_n$. If A has finite transcendence degree then any transcendence basis for A has this property. Conversely given a set with this property it contains a subset which is a transcendence basis for A over K (so in particular the transcendence degree of A over K is less than n).

Proof. If $\alpha_1, \ldots, \alpha_n$ is a transcendence basis for A over K then given $\alpha \in A$, $\alpha_1, \ldots, \alpha_n, \alpha$ must be algebraically dependent (or else it would be as algebraically independent set with more elements than a transcendence basis) so α depends algebraically on $\alpha_1, \ldots, \alpha_n$. Conversely if $\alpha_1, \ldots, \alpha_n$ is a subset of A on which all elements of A depend algebraically then we can pick a minimal subset of $\alpha_1, \ldots, \alpha_n$ with this property which, after renumbering, we can assume is $\alpha_1, \ldots, \alpha_r$. Then $\alpha_1, \ldots, \alpha_r$ are algebraically independent. (Otherwise there would be a first α_j such that $\alpha_1, \ldots, \alpha_j$ was algebraically dependent and then 2.2.15 and 2.2.16 show that every element of A depends algebraically on $\alpha_1, \ldots, \not{\alpha}_j, \ldots \alpha_r$). On the other hand by 2.2.17 no algebraically independent subset of A can have more than r elements. Hence $\alpha_1, \ldots, \alpha_r$ is a transcendence basis for A. ∎

2.2.20. Corollary. If $A = K[\alpha_1, \ldots, \alpha_n]$ then the transcendence degree of A over K is less than or equal to n.

2.2.21. Corollary. A subset $\alpha_1, \ldots, \alpha_n$ of A is a transcendence basis for A over K if and only if it has both of the following two properties:

(1) $\alpha_1, \ldots, \alpha_n$ are algebraically independent.

(2) Every element of A depends algebraically on $\alpha_1, \ldots, \alpha_r$.

If a subset satisfies (2) it has a subset which is a transcendence basis. If A has finite transcendence degree and a subset staisfies (1) then it is part of a transcendence basis. Thus if n is the transcendence degree of A over K and a subset $\alpha_1, \ldots, \alpha_n$ of A satisfies either (1) or (2) it also satisfies the other and hence is a transcendence basis for A over K.

2.2.22. Proposition. The field F of fractions of A has the same transcendence degree over K as does A. In fact if $\alpha_1, \ldots, \alpha_n$ is a transcendence basis for A over K it is also a transcendence basis for F over K.

Proof. Clearly the transcendence degree of F is at least as great as that of A, so suppose $\alpha_1, \ldots, \alpha_n$ is a transcendence basis for A. Then by 2.2.21 every element of A is algebraic over $K[\alpha_1, \ldots, \alpha_n]$. But by 2.2.9

the set of elements in F algebraic over $K[a_1, \ldots, a_n]$ is a field and so is all of F, i.e. every element of F depends algebraically on $a_1, \ldots, a_n$. ∎

2.2.23. Proposition. If the integral domain A has finite transcendence degree n over K, then n is the least integer with the property that any $n+1$ element subset of A is algebraically dependent over K.

Proof. Clearly every k-element subset of A is algebraically dependent over K if and only if there does not exist an algebraically independent set $a_1, \ldots, a_k$ in A. So this is just a logical rephrasing of 2.2.18. ∎

2.2.24. Lemma. Given $P(X_1, \ldots, X_n) \neq 0$ in $\mathbb{Z}[X_1, \ldots, X_n]$ there exist positive integers $m_1, \ldots, m_n$ such that $P(m_1, \ldots, m_n) \neq 0$.

Proof. Immediate from 1.2.15. ∎

2.2.25. Lemma. Given a finite set $J \subseteq \mathbb{Z}^n$ there exist positive integers $m_2, \ldots, m_n$ so that if $m = (1, m_2, \ldots, m_n)$ then the dot products $m \cdot j = j_1 + m_2 j_2 + \ldots + m_n j_n$ are distinct for different elements of J.

Proof. Let Δ denote the set of differences $j - j' = \delta$ with j, j' distinct elements of J. We want $m \cdot \delta \neq 0$ for any such δ. Thus it will suffice if $\prod_{\delta \in \Delta} (\delta_1 + m_2\delta_2 + \ldots + m_n\delta_n) \neq 0$ i.e. if $(m_2, \ldots, m_n)$ is not a root of the polynomial $\prod_{\delta \in \Delta} (\delta_1 + \delta_2 X_2 + \ldots + \delta_n X_n) = P(X_2, \ldots, X_n)$. But since $\delta \in \Delta$ clearly implies $\delta \neq 0$, $(\delta_1 + \delta_2 X_2 + \ldots + \delta_n X_n) \neq 0$ so $P(X_2, \ldots, X_n) \neq 0$

and 2.2.24 completes the proof. ∎

2.2.26. Theorem (Noether Normalization Lemma). Let K be any field and A any finitely generated K algebra. If A has transcendence degree r over K then there is a transcendence basis $z_1, \ldots, z_r$ for A over K such that every element of A is integral over the polynomial subalgebra $K[z_1, \ldots, z_r]$.

Proof. (Nagata). Let $x_1, \ldots, x_n$ be a set of generators for A. If $r = n$ then by 2.2.21 $x_1, \ldots, x_r$ is a transcendence basis for A over K and we may take $z_i = x_i$. Thus we can assume $r < n$ and proceed by induction on $n - r$. It will then suffice to show that there exist $y_2, \ldots, y_n$ in A such that every element of A is integral over $K[y_2, \ldots, y_n]$. For clearly $K[y_2, \ldots, y_n]$ must also have transcendence degree r over K, so by induction there will exist a transcendence base $z_1, \ldots, z_r$ for $K[y_2, \ldots, y_n]$ (and hence for A) such that every element of $K[y_2, \ldots, y_n]$ is integral over $K[z_1, \ldots, z_r]$, and hence by 2.2.6 such that every element of A is integral over $K[z_1, \ldots, z_r]$. Since $n > r$ $x_1, \ldots, x_n$ are algebraically dependent so there exists a polynomial $P(X_1, \ldots, X_n) \in K[X_1, \ldots, X_n]$ such that $P(x_1, \ldots, x_n) = 0$ but $P \neq 0$. Write $P(X_1, \ldots, X_n) = \sum_{j \in J} c_j X^j$ (where $X^j = X_1^{j_1} X_2^{j_2} \ldots X_n^{j_n}$ and the sum is over the finite non-empty set J of $j \in \mathbb{Z}^n$ such that $j_i \geq 0$ and $c_j \neq 0$). Let $m = (1, m_2, \ldots, m_n)$ where $m_2, \ldots, m_n$ are positive integers such that the dot products $m \cdot j = j_1 + m_2 j_2 + \ldots + m_n j_n$ are distinct for distinct elements j of J (2.2.25). Let $y_i = x_i - x_1^{m_i}$ $(i=2, \ldots n)$. We will show that x_1 satisfies a monic equation over

$K[y_2, \ldots, y_n]$ of degree d (equal to the maximum of $m \cdot j$ for $j \in J$). Thus x_1 is integral over $K[y_2, \ldots, y_n]$ and (since the elements integral over $K[y_2, \ldots, y_n]$ are ring by 2.2.6) it follows that $x_i = y_i + x_1^{m_i}$ is integral over $K[y_2, \ldots, y_n]$ and hence so is every element of $A = K[x_1, \ldots, x_n]$. If we substitute $X_i = Y_i + X_1^{m_i}$ in X^j $(i = 2, \ldots n)$ we get

$$X^j = X_1^{m \cdot j} + \sum_{\ell=0}^{m \cdot j - 1} A_\ell^j(Y_2, \ldots, Y_n) X_1^\ell .$$

It follows that if $\bar{j}$ is the unique element of J for which $m \cdot j = d$ then making the same substitution in $P(X_1, \ldots, X_n)$ gives

$$P(X_1, Y_2 + X_1^{m_1}, \ldots, Y_n + X_1^{m_n}) = c_{\bar{j}} X_1^d + \sum_{\ell=0}^{d-1} A_\ell(Y_2, \ldots, Y_n) X_1^\ell .$$

Substituting y_i for Y_i in the latter gives

$$Q(X_1) = P(X_1, y_2 + X_1^{m_1}, \ldots, y_n + X_1^{m_n}) \in K[y_2, \ldots, y_n][X_1]$$

Note that since $\bar{j} \in J$, $c_{\bar{j}} \neq 0$, so $c_{\bar{j}}$ is a unit of $K[y_2, \ldots, y_n]$ and so $Q(X_1)$ is essentially a monic polynomial in $K[y_2, \ldots, y_n][X_1]$. But since $y_i + x_1^{m_i} = x_i$ we have $Q(x_1) = P(x_1, x_2, \ldots, x_n) = 0$. ∎

2.2.27. <u>Proposition</u>. Let A have finite transcendence degree n over K and let $\mathcal{G}$ be a prime ideal of A. Then the transcendence degree m of $A/\mathcal{G}$ over K is less than or equal to n, and if m = n then $\mathcal{G} = (0)$.

<u>Proof</u>. Letting $\Pi : A \to A/\mathcal{G}$ denote the canonical homomorphism,

let $\beta_1, \ldots, \beta_r$ be algebraically independent elements of $A/\mathcal{I}$ and let $\beta_i = \prod(\alpha_i)$. Any polynomial in $K[X_1, \ldots, X_r]$ satisfied by $\alpha_1, \ldots, \alpha_r$ would also be satisfied by $\beta_1, \ldots, \beta_r$, so $\alpha_1, \ldots, \alpha_r$ are algebraically independent and hence $r \leq n$. This proves that $A/\mathcal{I}$ has finite transcendence degree m over K and that $m \leq n$. Now suppose $m = n$. Let $\beta_1, \ldots, \beta_n$ be a transcendence basis for $A/\mathcal{I}$, let $\alpha_1, \ldots, \alpha_n$ be as above (so that they are algebraically independent and hence a transcendence basis for A) and let $a \in \mathcal{I}$. Since a is algebraic over $K[\alpha_1, \ldots, \alpha_n]$ there is a non-zero polynomial $P(X)$ of least degree in $K[\alpha_1, \ldots, \alpha_n][X]$ satisfied by a, say $P(X) = \sum_{i=0}^{d} a_i(\alpha_1, \ldots, \alpha_n)X^i$. Now clearly $0 = \prod(P(a)) = \sum_{i=0}^{d} a_i(\beta_i, \ldots, \beta_n)\prod(a)^i$. But since $a \in \mathcal{I}$, $\prod(a) = 0$ so $a_0(\beta_1, \ldots, \beta_n) = 0$. But $\beta_1, \ldots, \beta_n$ are algebraically independent over K and hence a_0 must be the zero polynomial so $a_0(\alpha_1, \ldots, \alpha_n) = 0$. Then $0 = P(a) = a(\sum_{i=0}^{d-1} a_{i+1}(\alpha_1, \ldots, \alpha_n)a^i)$. Now $Q(X) = \sum_{i=0}^{d-1} a_{i+1}(\alpha_1, \ldots, \alpha_n)X^i$ has degree less than d = degree of P so $Q(a) \neq 0$, and it follows that $a = 0$. ∎

2.2.28. Corollary. Let A have finite transcendence degree n over K and let $\mathcal{I}$ be a prime ideal of A such that $A/\mathcal{I}$ has transcendence degree $n - 1$ over K. If J is a prime ideal of A with $J \subseteq \mathcal{I}$, then either $J = (0)$ or $J = \mathcal{I}$.

Proof. Applying 2.2.27 to the ideal I/J of A/J and recalling $A/\mathcal{I} \approx (A/J)/(I/J)$ it follows that A/J has transcendence degree $\geq n - 1$ and equality implies $J = I$. But again by 2.2.27 (with I replaced by J) inequality implies that A/J has transcendence degree n and that $J = (0)$.

■

2.2.29. Remark. We recall that $f(X) \in K[X]$ is called separable over K if it is irreducible over K and has no repeated roots in its splitting field. The latter condition is of course equivalent to $f(X)$ and $f'(X)$ being relatively prime in $K[X]$, and since $f(X)$ is irreducible and $f'(X)$ has lower degree, this will certainly be the case unless $f'(X) = 0$. If f has positive degree and K has characteristic zero then $f'(X) = 0$ is impossible, so every irreducible polynomial is separable. If K has characteristic p then $f'(X) = 0$ means simply that only powers of X divisible by p occur in $f(X)$. As we pointed out in the beginning of Section 2.0, if K is perfect than $f'(X) = 0$ implies $f(X)$ is the pth power of some $g(X) \in K[X]$ and so $f(X)$ is not irreducible.

An element α of an extension field of K which is algebraic over K is called separable over K if its minimal polynomial $f(X)$ is separable over K. By what has just been noted if K has characteristic zero or if K has prime characteristic and is perfect every α algebraic over K is separable over K.

2.2.30. Theorem of the Primitive Element.

Let K be an infinite field, $\alpha_1, \ldots, \alpha_n$ elements of an extension field which are algebraic over K and assume $\alpha_2, \ldots, \alpha_n$ are separable over K. Then there is an element θ of $K[\alpha_1, \ldots, \alpha_n]$, which is in fact a linear combination of $\alpha_1, \ldots, \alpha_n$, such that $K[\alpha_1, \ldots, \alpha_n] = K[\theta]$.

Proof. By an obvious induction argument it suffices to consider the case $n = 2$. So we take $\alpha_1 = \alpha$, $\alpha_2 = \beta$ and assume β is separable over K. In a splitting field for the minimal polynomials $f(X)$ and $g(X)$ for α and β respectively we have $f(X) = (X-\alpha_1)\dots(X-\alpha_r)$ and $g(X) = (X-\beta_1)\dots(X-\beta_r)$ where $\alpha = \alpha_1$, $\beta = \beta_1$ and $\beta_1, \dots, \beta_s$ are distinct. Let $\theta = \alpha + c\beta$ where $c \in K$ is any element not of the form $(\alpha_i-\alpha)/(\beta-\beta_j)$ $(i=1, \dots, r;$ $j=2, \dots, s)$. Since K is infinite such c certainly exist. It will suffice to show that $\beta \in K[\theta]$, for then since $\alpha = \theta - c\beta$, $\alpha \in K[\theta]$ follows and hence $K[\alpha, \beta] = K[\theta]$. Now the coefficients of the GCD of $g(X)$ and $f(\theta-cX)$ (considered as polynomials in the above splitting field) certainly lie in $K[\theta]$, for both $g(X)$ and $f(\theta-cX)$ are in $K[\theta][X]$ and by the Euclidean algorithm so is their GCD. But this GCD is just $X - \beta$. Indeed $g(\beta) = 0$ and $f(\theta-c\beta) = f(\alpha) = 0$ so $X - \beta$ divides $g(X)$ and $f(\theta-cX)$. Moreover since β is separable over K, β is a simple root of its minimal polynomial $g(X)$. And by choice of c no other root of $g(X)$ is a root of $f(\theta-cX)$. ■

2.2.31. Proposition. Assume that the integral domain A includes the field K and is a finitely generated K algebra, say $A = K[x_1, \dots, x_n]$, so that its field of fractions is $F = K(x_1, \dots, x_n)$. Let A have transcendence degree d over K. Then there exist elements $t_1, \dots, t_{d+1}$ in A with the following properties:

1) Each t_i is a linear combination of $x_1, \dots, x_n$ with coefficients in K.

2) $t_1, \dots, t_d$ are algebraically independent over K.

3) $K(t_1, \ldots, t_{d+1}) = K(x_1, \ldots, x_n) = F$; i.e. the subring $K[t_1, \ldots, t_d, t_{d+1}]$ of A has the same field of fractions as the whole ring A.

4) t_{d+1} is separable over $K(t_1, \ldots, t_d)$.

Proof. We can suppose $x_1, \ldots, x_d$ are algebraically independent. If $n = d$ we can take $t_1 = x_1, \ldots, t_d = x_d$, and $t_{d+1} = x_d$, so we proceed by induction on $n - d$. We suppose then that there exist $t_1, \ldots, t_{d+1}$ in $K[x_1, \ldots, x_{n-1}]$ such that each t_i is a linear combination of $x_1, \ldots, x_{n-1}$; $t_1, \ldots, t_d$ are algebraically independent, and $K(t_1, \ldots, t_{d+1}) = K(x_1, \ldots, x_{n-1})$. We do not assume t_{d+1} is necessarily separable over $K(t_1, \ldots, t_d)$ and as a first step show that if not then there is an $i \leq d$ such that $t_1, \ldots, t_{i-1}$, $t_{d+1}, t_{i+1}, \ldots, t_d$ are algebraically independent and t_i is separable over $K(t_1, \ldots, t_{i-1}, t_{d+1}, t_{i+1}, \ldots, t_d)$.

Since $t_1, \ldots, t_d$ are algebraically independent, $K[t_1, \ldots, t_d]$ is isomorphic to the ring of formal polynomials in d variables and in particular is a UFD (2.2.12). Since t_{d+1} is algebraic over $K[t_1, \ldots, t_d]$, by 2.2.13 the ideal of polynomials $P(t_1, \ldots, t_d, X_{d+1})$ in $K[t_1, \ldots, t_d][X_{d+1}]$ such that $P(t_1, \ldots, t_d, t_{d+1}) = 0$ is a principal ideal with generator $f(t_1, \ldots, t_d, X_{d+1})$ irreducible in $K[t_1, \ldots, t_d][X_{d+1}]$. By a remark at the beginning of Section 2.0, the formal partial derivative of $f(X_1, \ldots, X_{d+1})$ with respect to at least one of the variables $X_1, \ldots, X_{d+1}$ is not zero, say $f_i(X_1, \ldots, X_{d+1}) = (\partial/\partial X_i)f(X_1, \ldots, X_{d+1}) \neq 0$. If $i = d + 1$ then t_{d+1} is separable over $K[t_1, \ldots, t_d]$. Suppose $i \leq d$. Then f has positive degree with respect to X_i, and since $f(t_1, \ldots, t_i, \ldots, t_{d+1}) = 0$ t_i is algebraic over

$K[t_1, \ldots, t_{i-1}, t_{i+1}, \ldots, t_{d+1}]$. It follows that $t_1, \ldots, t_{i-1}, \ldots, t_{d+1}$ are algebraically independent, since otherwise $K[t_1, \ldots, t_{d+1}]$ would have transcendence degree less than d over K. Moreover t_i is separable over $K[t_1, \ldots, t_{i-1}, t_{i+1}, \ldots, t_{d+1}]$ since $f_i \neq 0$. Thus by reordering the t_i (interchanging t_i and t_{d+1}) we can suppose that t_{d+1} is separable over $K(t_1, \ldots, t_d)$ after all. Since x_n is algebraic over $K(x_1, \ldots, x_{n-1})$ and t_{d+1} is algebraic over $K(t_1, \ldots, t_d)$ we have $K(x_1, \ldots, x_n) = K(x_1, \ldots, x_{n-1})[K_n] = K(t_1, \ldots, t_{d+1})[x_n] = K(t_1, \ldots, t_d)[t_{d+1}, x_n]$. Then by 2.2.30 there is an element t'_{d+1} of $K(t_1, \ldots, t_d)[t_{d+1}, x_n]$ (which is in fact a linear combination of t_{d+1} and x_n, and so of $x_1, \ldots, x_n$, since t_{d+1} is a linear combination of $x_1, \ldots, x_{n-1}$) such that $K(x_1, \ldots, x_n) = K(t_1, \ldots, t_d)[t'_{d+1}] = K(t_1, \ldots, t_d, t'_{d+1})$.

∎

2.3. Dimension.

2.3.1. Definition. Let S be an affine algebraic space over K. The algebraic dimension of S over K, denoted by $DIM_K(S)$ (or simply DIM(S) when the identity of K is clear from the context) is a non-negative integer defined as follows:

If S is irreducible (so that by 1.7.18 the structure ring $\mathcal{P}(S)$ is an integral domain) then DIM(S) is the transcendence degree of $\mathcal{P}(S)$ over K.

In general DIM(S) is the maximum of DIM(V) where V is an irreducible component of S.

If all the irreducible components of S have the same dimension, say n, then we say that S is pure or has pure (algebraic) dimension n.

2.3.2. Remark. Suppose $\mathcal{P}(S)$ has n generators. If V is an irreducible subspace of S then $\mathcal{P}(S)$, being a quotient of $\mathcal{P}(S)$, also has n-generators; so by 2.2.20 $DIM(V) \leq n$. Thus $DIM(S) \leq n$.

2.3.3. Proposition. If V is a finite dimensional vector space over K then DIM(V) = dim(V), i.e. the algebraic dimension of V is its linear dimension. In particular $DIM(K^n) = n$.

Proof. If dim(V) = n then we have seen (cf. 2.0.8) that V is isomorphic as an algebraic space to K^n, so it suffices to consider this case. Now by 2.0.2 K^n is an irreducible algebraic variety and its structure ring is isomorphic to the algebra $K[X_1, \ldots, X_n]$ of polynomials in n-variables. But $X_1, \ldots, X_n$ is a transcendence basis for the latter and it follows that $DIM(K^n) = n$. ■

2.3.4. Proposition. If S is an algebraic space over K then $DIM(S)$ is the least integer n with the property that given any $n+1$ functions $f_1, \ldots, f_{n+1}$ in $\mathcal{P}(S)$ and any irreducible component T of S there exists $0 \neq P(X_1, \ldots, X_{n+1}) \in K[X_1, \ldots, X_{n+1}]$ such that $P(f_1, \ldots, f_{n+1})$ vanishes identically on T. The same is true if we allow T to be any irreducible subspace of S.

Proof. By 2.2.23 the first statement just says that $DIM(S)$ is the maximum of the transcendence degrees of $\mathcal{P}(T)$ for T an irreducible component of S, in agreement with 2.3.1. The second statement is trivial since (by 1.7.8) any irreducible subspace T of S is included in an irreducible component, and if the polynomial relation holds on the component it a fortiori holds on T. ∎

2.3.5. Corollary. If X is an algebraic space over K and Y is an algebraic subspace then $DIM(Y) \leq DIM(X)$. If X is irreducible and Y is a non-empty and Z-open in X then we have equality.

Proof. That we have inequality is immediate from the second statement of 2.3.4. If Y is a non-empty Z-open set in the irreducible space X then by 1.7.1 Y is Z-dense in X, so by 1.5.8 the restriction homomorphism $f \mapsto f|Y$ maps $\mathcal{P}(X)$ isomorphically into $\mathcal{P}(Y)$. Thus (cf. 1.7.18) $\mathcal{P}(Y)$ is also irreducible so that $DIM(Y)$ = transcendence degree of $\mathcal{P}(Y)$ = transcendence degree of $\mathcal{P}(X)$ = $DIM(X)$. ∎

2.3.6. Proposition. Let X be an irreducible algebraic space and let $Y \subseteq X$ be a Z-closed subspace of X. If $DIM(Y) = DIM(X)$ then $Y = X$.

Proof. We can assume Y is irreducible (otherwise apply the argument to an irreducible component of Y of maximal dimension). Since Y is Z-closed in X it will suffice to prove that Y is Z-dense in X, i.e. that any $f \in \mathcal{P}(X)$ which vanishes on Y vanishes on X. That is we must show that the ideal $\mathcal{I}$ of functions vanishing on Y is the zero ideal of $\mathcal{P}(X)$. Since Y is irreducible $\mathcal{I}$ is a prime ideal. Since by assumption $\mathcal{P}(X)$ and $\mathcal{P}(Y) \approx \mathcal{P}(X)/\mathcal{I}$ have the same transcendence degree over K the desired result follows from 2.2.2. $\blacksquare$

2.3.7. Corollary. If X is an irreducible algebraic variety and Y is any subvariety of X of the same dimension as X then $Y = X$.

2.3.8. Corollary. An algebraic variety of algebraic dimension zero is a finite set.

Proof. If X has algebraic dimension zero then each of its irreducible components has algebraic dimension zero, so it will suffice to assume that X is irreducible and prove that X is a point. But if $p \in X$ then $\{p\} \subseteq X$ and $DIM(\{p\}) = 0 = DIM(X)$, since clearly $\mathcal{P}(\{p\}) = K$. Hence $X = \{p\}$. $\blacksquare$

2.3.9. Definition. Let X be an algebraic space and let Y be an algebraic subspace of Y. If both X and Y are irreducible then we define

$$CODIM_X(Y) = DIM(X) - DIM(Y)$$

and we call $CODIM_X(Y)$ the (algebraic) codimension of Y in X. More generally if each irreducible component Y_0 of Y is included in a unique irreducible component X_0 of X and if $CODIM_{X_0}(Y_0)$ is the same for all such Y_0, then we define $CODIM_X(Y)$ to be this common value k of $CODIM_{X_0}(Y_0)$. In this case we say Y has pure codimension k in X.

2.3.10. Definition. Let X be an algebraic space and let H be a Z-closed, non-empty, proper subspace of X. We call H a hypersurface in X if the ideal $I(H)$ in $\mathcal{P}(X)$ of functions vanishing on H is principal (i.e. if there is an $F \in \mathcal{P}(X)$ such that $H = \{x \in X | F(x) = 0\}$ and $f \in \mathcal{P}(X)$ vanishes on H if and only if $f = gF$ for some $g \in \mathcal{P}(X)$).

2.3.11. Proposition. Let X be an irreducible algebraic space and suppose $\mathcal{P}(X)$ is a UFD (e.g. X might be K^n, cf. 2.2.12). Let H be a hypersurface in X and let $I(H) = (F)$. If $F = F_1F_2 \ldots F_k$ is the decomposition of F into irreducible factors then the F_i are distinct (i.e. F has no repeated factors), and if $H_i = \{x \in X | F_i(x) = 0\}$ then the H_i are irreducible hypersurfaces in X with $I(H_i) = (F_i)$, and in fact the H_i are precisely the irreducible components of H.

Proof. Since F is irreducible in $\mathcal{P}(X)$ if and only if the ideal (F) is prime, and $I(H) = (F)$ is prime if and only if H is irreducible, we see H is irreducible if and only if F is irreducible in $\mathcal{P}(X)$. We can suppose the factors of F so arranged that $F_1, \ldots, F_r$ are all the distinct irreducible factors which occur in F. Now at any $x \in H$ one of $F_1, \ldots, F_k$ at least vanish, hence at least one of $F_1, \ldots, F_r$ vanishes, i.e. $F_1F_2\ldots F_r$ vanishes on H and hence $F_1F_2\ldots F_k$ divides $F_1F_2\ldots F_r$, so $k = r$. As just remarked at any $x \in H$ one of the F_i vanishes, so H is the union of the Z-closed sets H_i. If f vanishes on H_1 then $fF_2\ldots F_k$ vanishes on H so $F_1F_2\ldots F_k$ divides $fF_2\ldots F_k$ and hence F_1 divides f. Thus $I(H_1) = (F_1)$ and similarly $I(H_i) = (F_i)$, so by the first remark in the proof H_i is an irreducible hypersurface. If H_1 were included in the union of $H_2, \ldots, H_k$, then $F_2F_3\ldots F_k$ would vanish on H and so would be divisible by $F_1F_2\ldots F_k$, which is impossible. Similarly none of the H_i can be included in the union of the others and so the H_i are the irreducible components of H. ∎

2.3.12. Corollary. Given X as in 2.3.11 and a Z-closed non-empty subset H of X, H is a hypersurface in X if and only if each irreducible component of H is a hypersurface in X.

Proof. One direction is part of 2.3.11. Suppose $H_1, \ldots, H_k$ are the irreducible components of H and $I(H_i) = (F_i)$. Note that by 2.3.11 the F_i are irreducible. Since $H_i = \{x \in X | F_i(x) = 0\}$ and H is the union of the H_i, $H = \{x \in X | F(x) = 0\}$ where $F = F_1F_2\ldots F_k$. If f vanishes on H it vanishes on each H_i, so each F_i divides f. Since the H_i and hence the F_i are distinct, $F = F_1F_2\ldots F_k$ divides f. Thus $(F) = I(H)$.

∎

2.3.13. Proposition. Let X be an irreducible algebraic space and let Y be a Z-closed subspace of X of pure codimension one. If X is an affine space, or more generally if $\mathcal{P}(X)$ is a UFD then Y is a hypersurface in X.

Proof. By 2.3.12 we can suppose that Y is irreducible. Let $f \in I(Y)$, $f \neq 0$, and let $f = P_1 P_2 \ldots P_k$ be the decomposition of f into irreducible factors. Since $I(Y)$ is prime at least one of these factors (call it P) is in $I(Y)$, so the non-zero prime ideal (P) is included on $I(Y)$. Since $\mathcal{P}(X)/I(Y) \simeq \mathcal{P}(Y)$ has transcendence degree $\mathrm{DIM}(X) - 1$ it follows from 2.2.28 that $I(Y) = (P)$. ∎

2.3.14. Proposition. If H is a hypersurface in K^n then H has pure codimension one in K^n. In fact if F is a generator for $I(H)$ then we can suppose (after a renumbering of the coordinate functions $X_1, \ldots, X_n$ of K^n) that $X_1, \ldots, X_{n-1}$, F is a transcendence basis for $\mathcal{P}(K^n)$, and it follows that $X_1, \ldots, X_{n-1}$ is a transcendence basis for $\mathcal{P}(K^n)$ modulo (F).

Proof. We note first that F cannot be algebraic over K, for this would imply that F is a constant polynomial and so either $H = \phi$ or $H = K^n$ which is impossible since H is a hypersurface. Then by 2.2.17 it follows that every element of $K[X_1, \ldots, X_n]$ depends algebraically on F and some $n - 1$ element subset of $\{X_1, \ldots, X_n\}$, which after renumbering we can assume is $X_1, \ldots, X_{n-1}$. Since $K[X_1, \ldots, X_n]$ has transcendence degree n it follows that $X_1, \ldots, X_{n-1}$, F is a transcendence basis for $K[X_1, \ldots, X_n]$. In particular they are algebraically independent, and hence $F(X_1, \ldots, X_n)$ cannot be a polynomial in $X_1, \ldots, X_{n-1}$, i.e. F must contain X_n essentially. It follows that any non-zero multiple of $F(X_1, \ldots, X_n)$ by an element of $K[X_1, \ldots, X_n]$ must contain X_n essentially, i.e. if $P(X_1, \ldots, X_{n-1}) \in (F)$ then $P = 0$, which says precisely that $X_1, \ldots, X_{n-1}$ are algebraically independent modulo (F). Since $\mathcal{P}(H) \simeq K[X_1, \ldots, X_n]/(F)$ and since we know H has algebraic dimension $\leq n - 1$ (by 2.3.5 and 2.3.6) it follows that $DIM(H) = n - 1$ and that the restrictions of $X_1, \ldots, X_{n-1}$ to H give a transcendence basis for $\mathcal{P}(H)$. ∎

2.3.15. Remark. We shall see later that when K is algebraically closed and F is any non-constant element of $K[X_1, \ldots, X_n]$ then $Y = \{x \in K^n \mid F(x_1, \ldots, x_n) = 0\}$ is always a hypersurface in K^n. If K is not algebraically closed this need not be the case. Indeed if $K = \mathbb{R}$ then every Z-closed subspace V of K^n can be so represented. For if I(V) is generated by $f_1, \ldots, f_m$ we need only take $F = f_1^2 + \ldots + f_m^2$. For example the origin in $\mathbb{R}^2$ is the zero set of the (clearly irreducible) polynomial $X^2 + Y^2$, but of course the ideal (X, Y) of $\{(0, 0)\}$ is not principal and $\{(0, 0)\}$ has dimension 0(codimension 2).

2.3.16. Proposition. If S_1 and S_2 are algebraic spaces over K then

$$\mathrm{DIM}(S_1 \times S_2) = \mathrm{DIM}(S_1) + \mathrm{DIM}(S_2) .$$

Proof. Recall (cf. 2.0.10 and 1.7.22) that $S_1 \times S_2$ is irreducible if and only if each factor is irreducible, so the irreducible component of $S_1 \times S_2$ are just the products $A \times B$ where A is an irreducible component of S_1 and B an irreducible component of S_2 (cf. 1.7.13). Thus we can assume that S_1 and S_2 are irreducible. Let $f_1, \ldots, f_n$ be a transcendence basis for $\mathcal{P}(S_1)$ and $g_1, \ldots, g_m$ a transcendence basis for $\mathcal{P}(S_2)$. Now in $\mathcal{P}(S_1 \times S_2) = \mathcal{P}(S_1) \otimes \mathcal{P}(S_2)$ every element of $\mathcal{P}(S_1)$ is algebraic over $K[f_1, \ldots, f_n]$ and a fortiori over $K[f_1, \ldots, f_n, g_1, \ldots, g_m]$. Similarly every element of $\mathcal{P}(S_2)$ is algebraic over $K[f_1, \ldots, f_n, g_1, \ldots, g_m]$. Since the set of elements of $\mathcal{P}(S_1 \times S_2)$ algebraic over $K[f_1, \ldots, f_n, g_1, \ldots, g_m]$ is a subring and includes $\mathcal{P}(S_1)$ and $\mathcal{P}(S_2)$, which generate $\mathcal{P}(S_1 \times S_2)$ as a ring it follows that every element of $\mathcal{P}(S_1 \times S_2)$ depends algebraically on $f_1, \ldots, f_n, g_1, \ldots, g_m$ and it remains only to see that the latter are algebraically independent and hence a transcendence basis. Suppose $P(f_1, \ldots, f_n, g_1, \ldots, g_m) = 0$ where $P(X_1, \ldots, X_n, Y_1, \ldots, Y_m) \in K[X_1, \ldots, X_n][Y_1, \ldots, Y_m]$, say $P = \sum_{\alpha} A_{\alpha}(X_1, \ldots, X_n) Y_1^{\alpha_1} \ldots Y_m^{\alpha_m}$. Then for any $s_1 \in S_1$ $P(f_1(s_1), \ldots, f_n(s_1), g_1, \ldots, g_m)$ vanishes identically on S_2. Since $g_1, \ldots, g_m$ are algebraically independent on S_2, each $A_{\alpha}(f_1(s_1), \ldots, f_n(s_1)) = 0$, i.e. $A_{\alpha}(f_1, \ldots, f_n)$ vanishes on S_1, and since $f_1, \ldots, f_n$ are algebraically independent on S_1 $A_{\alpha}(X_1, \ldots, X_n) = 0$ so $P(X_1, \ldots, X_n, Y_1, \ldots, Y_m) = 0$. ∎

2.4. Simple Points and Nonsingular Spaces

2.4.0 Remark. Let V be a finite dimensional vector space over K, regarded as an algebraic space. If L is a linear isomorphism of V with K^n then L is also an isomorphism of algebraic spaces. From this fact and 1.9.28 we see that for each $v_0 \in V$ we have a canonical isomorphism $v \mapsto \partial^v_{v_0}$ of V with $(TV)_{v_0}$, the tangent space to V at v_0. Since the algebraic dimension $\mathrm{DIM}(V)$ of V is equal to its linear dimension it then follows that $\dim(TV)_{v_0} = \mathrm{DIM}(V)$ for all $v_0 \in V$. For completeness we recall the definition of $\partial^v_{v_0}$. Given $v \in V$ and $f \in \mathcal{P}(V)$ there is a unique polynomial Q^v_f with coefficients in $\mathcal{P}(V)$ such that for all $t \in K$ and $x \in V$, $f(x+tv) = Q^v_f(t)(x)$. Clearly Q^v_f has constant term f and the coefficient of T in $Q^v_f(T)$ is thus $((f(x+Tv)-f(x))/T)_{T=0} = \partial^v f \in \mathcal{P}(V)$. The map $\partial^v : \mathcal{P}(V) \to \mathcal{P}(V)$ is a vector field on V, i.e. a derivation of the K-algebra $\mathcal{P}(V)$, called the directional derivative in the direction v (cf. 1.9.18). Then $\partial^v_{v_0} : \mathcal{P}(V) \to K$ is just ∂^v composed with evaluation at v_0 (i.e. $\partial^v_{v_0} f = (\partial^v f)(v_0) = [(f(v_0+Tv)-f(v_0))/T]_{T=0}$.

If $e_1, \ldots, e_n$ is a basis for V and $X_1, \ldots, X_n$ is the dual basis for V^* then $\mathcal{P}(V)$ is the polynomial algebra $K[X_1, \ldots, X_n]$ and ∂^{e_i} is the formal partial derivative $\partial/\partial X_i$. At each $v \in V$ $(dX_1)_v, \ldots, (dX_n)_v$ is a basis for $(T^*V)_v$ and $\partial^{e_1}_v, \ldots, \partial^{e_n}_v$ is the dual basis, so that for any P in $\mathcal{P}(V)$ we have $(dP)_v = \sum_{i=1}^n (\partial P/\partial X_i)_v (dX_i)_v$, or symbolically $dP = \sum_{i=1}^n (\partial P/\partial X_i) dX_i$.

2.4.1. Proposition. For any algebraic space S, the set of points s of S where $(TS)_s$ has dimension greater than or equal to some integers k is a Z-closed subset of S.

Proof. We can assume S is an algebraic subspace of some K^n and let $I = I(S)$ denote the ideal of g in $\mathcal{P}(K^n) = K[X_1, \ldots, X_n]$ vanishing on S. Denote by dI_s the set $\{dg_s \mid g \in I\} \subseteq (T^*S)_s$. Then (by 1.7.13) $(TS)_s$, considered as a subspace of $(TK^n)_s$, is just the annihilator of dI_s, while (by 1.9.14) if $g_1, \ldots, g_m$ generate I then dI_s is spanned by the $(dg_i)_s$. Thus $(TS)_s$ has dimension $\geq k$ if and only if dI_s has dimension $\leq r = n - k$. This in turn is the case if and only if for all $1 \leq \mu_1 < \ldots < \mu_{r+1} \leq m$, $(dg_{\mu_1})_s, \ldots, (dg_{\mu_{r+1}})_s$ are linearly dependent, or what is the same, the Grassman product $(dg_{\mu_1})_s \wedge \cdots \wedge (dg_{\mu_{r+1}})_s$ is zero. Now since $dg = \sum_{i=1}^{n} (\partial g/\partial X_i) dX_i$, $dg_{\mu_1} \wedge \cdots \wedge dg_{\mu_{r+1}} = \sum_{\nu} A^{\mu}_{\nu} dX_{\nu_1} \wedge \cdots \wedge dX_{\nu_{r+1}}$ where A^{μ}_{ν} is the Jacobian determinant $\partial(g_{\mu_1}, \ldots, g_{\mu_{r+1}})/\partial(X_{\nu_1}, \ldots, X_{\nu_{r+1}})$ (which is a polynomial because the g_i are polynomials). The sum is over all $1 \leq \nu_1 < \ldots < \nu_{r+1} \leq m$. Since $(dX_1)_s, \ldots, (dX_n)_s$ is a basis for $(T^*K^n)_s$, these $(dX_{\nu_1})_s \wedge \cdots \wedge (dX_{\nu_{r+1}})_s$ are a basis for $\Lambda^{r+1}(T^*K^n)_s$, so dI_s has dimension $\leq r$ precisely where all the polynomials A^{μ}_{ν} vanish on S. ∎

2.4.2. Definition. Let S be an algebraic space over K. A point s of S is called a simple point of S if S is irreducible and $\dim (TS)_s = DIM(S)$; more generally if S is not necessarily irreducible then s is called a simple point of S if it belongs to exactly one irreducible component of S and is a simple point of that component. S is called non-singular or smooth if each of its points is simple. The complement in S of the set of simple points is called the singular set of S and is denoted by $\Sigma(S)$. The set $S - \Sigma(S)$ of simple points of S will be denoted by S_{NS}.

2.4.3. Remark. Note that it is immediate from the definition that an algebraic space S is smooth if and only if it is the disjoint union of its irreducible components and each of them is smooth.

If V is a finite dimensional vector space over K then since V is an irreducible algebraic variety it follows from 2.4.0 that every point of V is simple and hence that V is smooth (or non-singular).

2.4.4. Proposition. Let X be an algebraic space and let X' be an open algebraic subspace of X. A point x of X' is a simple point of X' if and only if it is a simple point of X, i.e. $X'_{NS} = X' \cap X_{NS}$.

Proof. First suppose X is irreducible. Then X' is a non-empty Z-open (and hence Z-dense) subspace of X, so that the restriction map $f \mapsto f|X'$ is an isomorphism $\mathcal{P}(X) \to \mathcal{P}(X')$. Since DIM(X) is the transcendence degree of $\mathcal{P}(X)$ and TX_x is the space of point derivations of $\mathcal{P}(X)$ at x, it follows that $DIM(X) = \dim TX_x$ if and only if $DIM(X') = \dim TX'_x$.

In general let $X_1, \ldots, X_n$ be the irreducible components of X which meet X', so that by 1.7.15 the $X'_i = X_i \cap X'$ are the irreducible components of X'. Thus $x \in X'$ belongs to a unique component of X' if and only if it belongs to a unique component of X. Moreover, since X'_x is open in X_i, a point x of X'_i is a simple point of X'_i if and only if it is a simple point of X_i. ■

2.4.5. Corollary. Let X be an algebraic space and let Y be the subspace of X consisting of points belonging to a unique irreducible component of X. Then Y is an open, dense algebraic subspace of X whose irreducible and connected components are the intersections of Y with the irreducible components of X. Moreover $Y_{NS} = X_{NS}$.

Proof. The first conclusion is just 4) of 1.7.15 and the second then follows immediately from the proposition. ■

2.4.6. Proposition. If S is an irreducible hypersurface in K^n then $\Sigma(S)$, the singular set of S, is a proper subvariety of S and hence the set S_{NS} of simple points of S is a Z-dense open set.

Proof. By definition the ideal I(S) of functions in $K[X_1, \ldots, X_n] = \mathcal{P}(K^n)$ vanishing on S is a nonzero, proper, principal ideal (f) (cf. 2.3.10), and since S is irreducible f is a nonconstant irreducible element of $K[X_1, \ldots, X_n]$ (cf. 2.3.11) so that by 2.0 the formal partial derivatives $\partial f/\partial X_i$ cannot all be the zero polynomial. Now if $\partial f/\partial X_i$ is not zero it cannot be a multiple of f, since considered as a polynomial in X_i it has smaller degree than f. Since I(S) = (f) it follows that not all the $\partial f/\partial X_i$ can vanish identically on S so that $S' = \{x \in S \mid (\partial f/\partial X_i)(x) = 0, i=1, \ldots, n\}$ is a (possibly empty) proper subvariety of S. We shall show that in fact $S' = \Sigma(S)$. Since $DIM(S) = n - 1$ by 2.3.14 what we must show is that if $x \in S$ then $\dim(TS)_x = n - 1$ if and only if $x \in S'$. Now by 1.9.14 $(TS)_x = \{D \in (TK^n)_x \mid df_x(D) = 0\}$. Since $\dim(TK^n)_x = n$, $(TS)_x$ has dimension n where $df_x = 0$ and has dimension $n - 1$ where $df_x \neq 0$, so we must show that $S' = \{x \in S \mid df_x = 0\}$. Since $df = \sum_{i=1}^{n} (\partial f/\partial X_i) dX_i$ and $(dX_i)_x$ is a basis for $(TK^n)_x$ this is clear. ∎

2.4.7. Definition. Let X and Y be irreducible algebraic spaces and let $\mathcal{F}(X)$ and $\mathcal{F}(Y)$ denote respectively the fields of quotients of the two integral domains $\mathcal{P}(X)$ and $\mathcal{P}(Y)$. Let $\varphi : X \to Y$ be a morphism of algebraic spaces such that $\varphi(X)$ is Z-dense in Y, so that $\varphi^* : \mathcal{P}(Y) \to \mathcal{P}(X)$ is injective and therefore extends to an embedding of fields $\varphi^* : \mathcal{F}(Y) \to \mathcal{F}(X)$. If this latter map is actually on isomorphism of $\mathcal{F}(Y)$ with $\mathcal{F}(X)$ then we say that φ is a birational equivalence of X with Y.

2.4.8. Proposition. Let $\varphi : X \to Y$ be a birational equivalence of irreducible algebraic spaces over K. There is a Z-dense open set U of X such that:

1) φ maps U one-to-one onto a Z-dense subset $\varphi(U)$ of Y.

2) For each x in U the induced map of local rings (cf. 1.10.4)

$$\mathcal{O}_{\varphi, x} : \mathcal{O}_{Y, \varphi(x)} \to \mathcal{O}_{X, x}$$

is an isomorphism, so that (cf. 1.10.13)

$$T\varphi_x : TX_x \to TY_{\varphi(x)}$$

is a linear isomorphism.

Proof. Let $\xi = (\xi_1, \ldots, \xi_n)$ and $\eta = (\eta_1, \ldots, \eta_n)$ be generating points for X and Y respectively (cf. 2.1.1). Recall that $\mathcal{P}(X)$ is canonically isomorphic to $K[X_1, \ldots, X_n]/\mathcal{J}^{\xi}$ so that we may take the field $\mathcal{F}(X)$ of quotients to be represented by quotients of elements of $K[X_1, \ldots, X_n]$ with denominators not in $\mathcal{J}^{\xi}$. Similarly $\mathcal{F}(Y)$ is represented by quotients of elements of $K[Y_1, \ldots, Y_m]$ with denominator not in $\mathcal{J}^{\eta}$. By assumption $f \mapsto f \circ \varphi$ is a monomorphism $\varphi^* : \mathcal{P}(Y) \to \mathcal{P}(X)$ which extends to an isomorphism $\varphi^* : \mathcal{F}(Y) \to \mathcal{F}(X)$. Define $h_i \in \mathcal{F}(Y)$, $i = 1, \ldots, n$, by $\varphi^*(h_i) = \xi_i$ and let $h_i = P_i(\eta_1, \ldots, \eta_m)/Q_i(\eta_1, \ldots, \eta_m)$ where $P_i, Q_i \in K[Y_1, \ldots, Y_m]$ and $Q_i \notin \mathcal{J}^{\eta}$. Since $\mathcal{J}^{\eta}$ is a prime ideal (because Y is irreducible) $Q = Q_1 Q_2 \ldots Q_n \notin \mathcal{J}^{\eta}$ so $q = Q(\eta_1, \ldots, \eta_m)$ is not the zero element of $\mathcal{P}(Y)$ and $\mathcal{O} = \{ y \in Y \mid q(y) \neq 0\}$ is a non-empty (and hence Z-dense) open set in Y. Since by assumption $\varphi(X)$ is also Z-dense in Y, $U = \varphi^{-1}(\mathcal{O})$ is a Z-dense open set of X and $\varphi(U)$ is dense in Y.

We note next that for $P \in K[X_1, \ldots, X_n]$ $\varphi^*(P(h_1, \ldots, h_n)) = P(\varphi^* h_1, \ldots, \varphi^* h_n) = P(\xi_1, \ldots, \xi_n)$. So that since $\varphi^* : \mathcal{F}(Y) \to \mathcal{F}(X)$ is an isomorphism, $P(h_1, \ldots, h_n) = 0$ if and only if $\varphi^*(P(h_1, \ldots, h_n)) = 0$, if and only if $P \in \mathcal{J}^{\xi}$. It follows that the map $P \mapsto P(h_1, \ldots, h_n)$ induces a well defined monomorphism of $\mathcal{P}(X) = K[X_1, \ldots, X_n]/\mathcal{J}^{\xi}$ into $\mathcal{F}(Y)$ which extends to a field embedding $\theta : \mathcal{F}(X) \to \mathcal{F}(Y)$, namely $P(\xi_1, \ldots, \xi_n)/Q(\xi_1, \ldots, \xi_n) \mapsto P(h_1, \ldots, h_n)/Q(h_1, \ldots, h_n)$. Now $\theta(\xi_i) = h_i = \varphi^{*-1}(\xi_i)$ and since the ξ_i generate $\mathcal{P}(X)$ as an algebra (and hence generate $\mathcal{F}(X)$ as a field) it follows that $\theta = \varphi^{*-1}$. Since $\xi_i = \varphi^*(h_i)$ and

$h_i = P_i(\eta)/Q_i(\eta)$ it follows that in $\mathcal{P}(X)$ we have the equations $\xi_i Q_i(\varphi^*(\eta_1), \ldots, \varphi^*(\eta_m)) = P_i(\varphi^*(\eta_1), \ldots, \varphi^*(\eta_m))$. If we take x in U and let $x_i = \xi_i(x)$, $y = \varphi(x)$ and $y_j = \eta_j(y)$ then since $\varphi^*(\eta_j) = \eta_j \circ \varphi_j$ we get $x_i Q_i(y_1, \ldots, y_m) = P_i(y_1, \ldots, y_n)$. Now $y = \varphi(x) \in \varphi(U) \subseteq \mathcal{O}$ so $0 \neq q(y) = \prod_i Q_i(y_1, \ldots, y_m)$ so $Q_i(y_1, \ldots, y_m) \neq 0$ and hence $\xi_i(x) = x_i = P_i(y_1, \ldots, y_m)/Q_i(y_1, \ldots, y_m)$. Thus $\xi_i(x)$ is determined by the $y_i = \eta_i(\varphi(x))$ and hence by $\varphi(x)$. Since the $\xi_i(x)$ in turn determine x (cf. 2.1.2) it follows that φ is one-to-one on U. Since X is irreducible it follows from 1.10.3 that we can identify $\mathcal{O}_{X,x}$ with the subring of $\mathcal{F}(X)$ consisting of quotients f/g, $f, g \in \mathcal{P}(X)$ and $g(x) \neq 0$, and similarly $\mathcal{O}_{Y,\varphi(x)}$ is the corresponding subring of $\mathcal{F}(Y)$. Since $\varphi^* : \mathcal{F}(Y) \to \mathcal{F}(X)$ is an isomorphism and clearly induces $\mathcal{O}_{\varphi,x} : \mathcal{O}_{Y,\varphi(x)} \to \mathcal{O}_{X,x}$ the latter is certainly injective and it remains only to show that (still assuming $x \in U$ so $y = \varphi(x) \in \mathcal{O}$) it is surjective. Given $P(\xi_1, \ldots, \xi_n)/Q(\xi_1, \ldots, \xi_n)$ with $Q(x_1, \ldots, x_n) \neq 0$ we must check that its image $P(h_1, \ldots, h_n)/Q(h_1, \ldots, h_n)$ under $\theta = \varphi^{*-1}$ is in $\mathcal{O}_{Y,\varphi(x)}$. Now while $Q(h_1, \ldots, h_n)$ does not vanish at $y = \varphi(x)$ (since $h_i(y) = h_i(\varphi(x)) = \varphi^*(h_i)(x) = \xi_i(x) = x_i$), $Q(h_1, \ldots, h_n)$ is not necessarily in $\mathcal{P}(Y)$. However recall that $h_i = P_i(\eta_1, \ldots, \eta_m)/Q_i(\eta_1, \ldots, \eta_m)$ so that multiplication by a suitable power of $q = \prod_{i=1}^{n} Q_i(\eta_1, \ldots, \eta_m)$ will clear both $P(h_1, \ldots, h_n)$ and $Q(h_1, \ldots, h_n)$ of denominators. Moreover since $q(y) \neq 0$ we will get thereby a representation of $P(h_1, \ldots, h_n)/Q(h_1, \ldots, h_n)$ in the form f/g where $f, g \in \mathcal{P}(Y)$ and $g(y) \neq 0$, proving that it is in $\mathcal{O}_{Y,y}$ as required. ■

2.4.9. Proposition. Let X be an irreducible algebraic space over K of algebraic dimension d. Then there is an irreducible hypersurface S in K^{d+1} for which there exists a birational equivalence $\varphi : X \to S$.

Proof. Let $\xi = (\xi_1, \ldots, \xi_n)$ be a generating point for X so that $\mathcal{P}(X) = K[\xi_1, \ldots, \xi_n]$, and hence $\mathcal{F}(X)$, the field of fractions of $\mathcal{P}(X)$ is $K(\xi_1, \ldots, \xi_n)$. By 2.2.31 we can find elements $t_1, \ldots, t_{d+1}$ in $\mathcal{P}(X)$ such that $t_1, \ldots, t_d$ are algebraically independent over K (so that $t_1, \ldots, t_d$ is a transcendance basis for $\mathcal{P}(X)$, and $K[t_1, \ldots, t_d] \subseteq \mathcal{P}(X)$ is a polynomial ring) and such that $\mathcal{F}(X) = K(t_1, \ldots, t_{d+1})$. We can suppose $t_{d+1} \neq 0$ (otherwise replace it by t_d). Let I denote the ideal of polynomials $P(Y_1, \ldots, Y_{d+1})$ in $K[Y_1, \ldots, Y_{d+1}]$ such that $P(t_1, \ldots, t_{d+1}) = 0$. Since $t_1, \ldots, t_d$ are algebraically independent, $f(Y_1, \ldots, Y_{d+1}) \mapsto f(t_1, \ldots, t_d, Z)$ is an isomorphism of $K[Y_1, \ldots, Y_{d+1}]$ with $K[t_1, \ldots, t_d][Z]$ and since $K[t_1, \ldots, t_n]$ is a polynomial ring and hence a UFD it follows from 2.2.13 that I is a principal prime ideal in $K[Y_1, \ldots, Y_{d+1}]$, say $I = (f)$. Let S be the subvariety of K^{d+1} defined by $S = \{y \in K^{d+1} | f(y) = 0\}$ and let $\eta_i = Y_i | S$. If we define a polynomial map $\varphi : X \to K^{d+1}$ by $x \mapsto (t_1(x), \ldots, t_{d+1}(x))$, then since $f(t_1(x), \ldots, t_{d+1}(x)) = 0$ by definition of f, it follows that in fact $\varphi : X \to S$. Moreover $\varphi^*(\eta_i) = \eta_i \circ \varphi = Y_i \circ \varphi = t_i$. If $P \in I(S)$ then $P(\eta_1, \ldots, \eta_{d+1}) = 0$ hence $0 = \varphi^* P(\eta_1, \ldots, \eta_{d+1}) = P(\varphi^* \eta_1, \ldots, \varphi^* \eta_{d+1}) = P(t_1, \ldots, t_{d+1})$ so $P \in I = (f)$. Since conversely $f \in I(S)$ and hence $(f) \subseteq I(S)$ we have $(f) = I(S)$, so $I(S)$ is a principal prime ideal of $K[Y_1, \ldots, Y_{d+1}]$ and hence by 2.3.10 and 2.3.11 S is an irreducible hypersurface in K^{d+1}. As we have just seen, if $\varphi^* P(\eta_1, \ldots, \eta_{d+1}) = 0$ then $P \in I(S)$ so $P(\eta_1, \ldots, \eta_{d+1}) = 0$ i.e. $\varphi^* : \mathcal{P}(S) \to \mathcal{P}(X)$ is injective. Since $\varphi^*(\eta_i) = t_i$ and $\mathcal{F}(X) = K(t_1, \ldots, t_{d+1})$ it follows that φ^* extends to an isomorphism of the field of quotients $\mathcal{F}(S)$ of $\mathcal{P}(S)$ with $\mathcal{F}(X)$. ■

2.4.10. Theorem. If X is an algebraic space over K then $\Sigma(S)$, the singular set of X, is a Z-closed nowhere dense subset of X. Equivalently X_{NS} is an open, dense, nonsingular algebraic subspace of X.

Proof. By 2.4.5 we can assume that the irreducible components of X are disjoint. Since each component is then open in X it suffices to consider the case that X is irreducible. If $\mathrm{DIM}(X) = d$ then by 2.4.9 there is an irreducible hypersurface S in K^{d+1} and a birational equivalence $\varphi : X \to S$. By 2.4.6. S_{NS} is an open dense subset of S. On the other hand by 2.4.8 there is an open dense subset U of X such that $\varphi(U)$ is dense in S and for $x \in U$, $T\varphi_x : TX_x \to TS_{\varphi(x)}$ is an isomorphism. Now since S_{NS} is open in S, $\varphi(U) \cap S_{NS}$ is non-empty and hence $U' = U \cap \varphi^{-1}(S_{NS})$ is a non-empty (and hence Z-dense) Z-open subset of X. For $x \in U'$ we have $y = \varphi(x) \in S_{NS}$ so that $\dim TX_x = \dim TS_y = \mathrm{DIM}(S) = d = \mathrm{DIM}(X)$. Let $\delta = \min\{\dim TX_x \mid x \in X\}$ and let $\mathcal{O} = \{x \in X \mid \dim TX_x = \delta\} = \{x \in X \mid \dim TX_x < \delta + 1\}$, so that by 2.4.1 $\mathcal{O}$ is a Z-open subset of X which is trivially non-empty and hence Z-dense. It follows that $U' \cap \mathcal{O}$ is non-empty. Since for x in $U' \cap \mathcal{O}$ we have $d = \dim TX_x = \delta$ it follows that $\mathcal{O} = \{x \in X \mid \dim TX_x = \mathrm{DIM}(X)\} = X_{NS}$. ■

2.4.11. <u>Corollary.</u> If X is an irreducible algebraic space over K then for all $x \in X$ $\dim(TX_x) \geq \mathrm{DIM}(X)$.

2.4.12. <u>Corollary.</u> Let X be an irreducible algebraic subspace of K^N and let $I = \{f \in \mathcal{P}(K^n) \mid (f|X) = 0\}$. Then for $x \in X$, $\dim(\{df_x \mid f \in I\}) \leq N - \mathrm{DIM}(X)$, with equality if and only if x belongs to the Z-open and dense subspace X_{NS} of X.

<u>Proof.</u> Immediate from 1.9.13 and 2.4.11. ■

2.4.13. <u>Proposition.</u> Let X be an irreducible algebraic subvariety of $\mathbb{R}^m$ of algebraic dimension d and let x_0 be a simple point of X. Choose $f_1, \ldots, f_\delta$ in $I = \{f \in \mathcal{P}(\mathbb{R}^m) \mid (f|X) = 0\}$ such that $(df_1)_{x_0}, \ldots, (df_\delta)_{x_0}$ is a basis for $\{df_{x_0} \mid f \in I\}$ (so that by 2.4.12 $\delta = m - d$) and choose $f_{\delta+1}, \ldots, f_m$, linear functionals on $\mathbb{R}^m$ so that $(df_1)_{x_0}, \ldots, (df_n)_{x_0}$ is a basis for $T^*\mathbb{R}^m_{x_0}$. Thus by the C^ω inverse function theorem $f_1, \ldots, f_m$ is a C^ω coordinate system for $\mathbb{R}^m$ near x_0, i.e. there is an $\epsilon > 0$ such that $x \mapsto (f_1(x), \ldots, f_m(x))$ is a C^ω diffeomorphism of a neighborhood $\mathcal{O}$ of x_0 in $\mathbb{R}^m$ onto $\{y \in \mathbb{R}^m \mid |y_i - f_i(x_0)| < \epsilon\}$. Moreover if ϵ is sufficiently small then $S = \{x \in \mathcal{O} \mid f_1(x) = \ldots = f_\delta(x) = 0\}$, $\mathcal{O} \cap X$ and $\mathcal{O} \cap X_{NS}$ are all equal and in particular $x \mapsto (f_{\delta+1}(x), \ldots, f_m(x))$ maps $\mathcal{O} \cap X_{NS}$ one-to-one onto $\{y \in \mathbb{R}^d \mid |y_i - f_i(x_0)| < \epsilon\}$.

<u>Proof.</u> Since $\Sigma(X)$ is Z-closed in X and a fortiori closed in the W-topology (cf. 1.5.14 and 1.5.15) it is clear that $\mathcal{O} \cap X = \mathcal{O} \cap X_{NS}$ if ϵ is

sufficiently small. Since $f_1, \ldots, f_\delta$ belong to I it is trivial that $\mathcal{O} \cap X_{NS} \subseteq S$. Since X is a subvariety of $\mathbb{R}^m$, $X = \{x \in \mathbb{R}^m \mid g(x) = 0 \text{ for all } g \in I\}$. Thus to prove the reverse inclusion $S \subseteq \mathcal{O} \cap X_{NS}$ it will suffice to show that any $g \in I$ vanishes identically on S. Now S is a connected, C^ω manifold and $f_{\delta+1}, \ldots, f_m$ is a global C^ω coordinate system for S. Hence it will suffice to show that if $g \in I$ then all the partial derivatives of $g|S$ at x_0 with respect to these coordinates vanish. Now the set of x where $(df_1)_x, \ldots, (df_\delta)_x$ are linearly independent is clearly open so that since it contains x_0 it will include $\mathcal{O}$ if ϵ is small enough. Then since $\mathcal{O} \cap X = \mathcal{O} \cap X_{NS}$ it follows from 2.4.13 that dg_x depends linearly on $(df_1)_x, \ldots, (df_\delta)_x$ at all points of $\mathcal{O} \cap X$. From lemma b of 1.6.13 we see, using the notation of that section, that $\Phi_\mu(g) \in I$ so by lemma c of that section, and the rule for differentiating a product, it follows by induction that any partial derivative $(\partial^r g/\partial f_{j_1} \ldots \partial f_{j_r})$ with all $j_i > \delta$ can be written in $\mathcal{O}$ as a sum of terms Ah where $A \in C^\omega(\mathcal{O}, \mathbb{R})$ and $h \in I$. In particular all such partial derivatives vanish at points of $\mathcal{O} \cap X$, and hence they all vanish at x_0. ∎

2.4.14. Corollary. X_{NS} is a regularly embedded (but not necessarily closed) d-dimensional C^ω submanifold of $\mathbb{R}^m$.

Proof. Trivial ∎

2.4.15. Theorem. Let V be a real algebraic variety and let M denote V_{NS} with its W-topology. There is a uniquely determined structure of C^ω manifold for M such that whenever $\xi = (\xi_1, \ldots, \xi_m)$ is a generating point for V $x \mapsto (\xi_1(x), \ldots, \xi_m(x))$ is a C^ω diffeomorphism of M with a regularly embedded C^ω submanifold of $\mathbb{R}^m$. The different connected components of M will in general have different dimensions; if $x \in M$ then the dimension of the connected component of M which contains x is the algebraic dimension of the irreducible component of V containing x.

Proof. Since V_{NS} is the finite disjoint union of its irreducible components each of which is Z-open and closed (and a fortiori W-open and closed) we can assume V_{NS} and hence V is irreducible. The uniqueness of a C^ω structure with the given property is obvious. Let $\xi = (\xi_1, \ldots, \xi_m)$ be a generating point for V. Then $x \mapsto (\xi_1(x), \ldots, \xi_m(x))$ is an isomorphism of V with an algebraic subvariety X of $\mathbb{R}^m$ (cf. 2.1.3). Since V_{NS} is mapped one-to-one onto X_{NS} we can give M the structure of a C^ω manifold by declaring $x \mapsto \xi(x)$ to be a C^ω diffeomorphism of M with the C^ω submanifold X_{NS} of $\mathbb{R}^m$ (cf. 2.4.14). If $\eta = (\eta_1, \ldots, \eta_n)$ is another generating point for V then each η_i is a polynomial in $\xi_1, \ldots, \xi_m$ and it follows that $x \mapsto \eta(x)$ is a C^ω map. Interchanging the roles of ξ and η it follows that the C^ω structures induced by ξ and η are the same. ∎

2.4.16. Corollary. If V is a nonsingular, irreducible real algebraic variety of algebraic dimension d then there is a unique structure of C^ω manifold of dimension d for (the underlying point set of) V such that whenever $\xi = (\xi_1, \ldots, \xi_m)$ is a generating point for V, $E^\xi : V \to \mathbb{R}^m$ is a proper, C^ω embedding of V on a closed, regularly embedded C^ω submanifold of $\mathbb{R}^m$.

2.4.17. Remark. The C^ω manifolds that arise as above might seem at first glance to be special and exceptional. It is however a remarkable fact, proved by Nash and Tognoli that every compact, connected C^ω manifold arises as in 2.4.16 from some irreducible nonsingular real algebraic variety V. The proof of this and related facts will be given later.

REFERENCES

1. Artin, M., "Tognoli's proof of the Nash conjecture", unpublished.

2. Artin, M., and Mazur, B., "On periodic points", Annals of Math. 81 (1965), 82-99.

3. Bialnicki-Birula, A., and Rosenlicht, M., "Injective morphisms of real algebraic varieties", Proc. AMS, 13 (1962), 200-203.

4. Borel, A., Linear Algebraic Groups, Benjamin, New York, 1969.

5. Cartan, H., "Variétés analytiques réeles et variétés analytiques complexes", Bull. Soc. Math. France, 85 (1957), 77-99.

6. Chevalley, C., Theory of Lie Groups, Princeton University Press, Princeton, 1946.

7. Chillingworth, D. R., and Hubbard, J. A., "Note on non-rigid Nash structures", Bull. AMS, 77 (1971), 429-431.

8. de Rham, G., Variétés Différentiables, Herman, Paris, 1955.

9. Dieudonné, J., "Algebraic geometry", Advances in Math., 3 (1969), 231-321.

10. Fogarty, J., Invariant Theory, Benjamin, New York, 1969.

11. Grauert, H., "On Levi's problem and the embedding of real analytic manifolds", Annals of Math., 68 (1958), 460-472.

12. Gunning, R., and Rossi, H., Analytic Functions of Several Complex Variables, Prentice Hall, Englewood, N. J., 1965.

13. Hartshorne, R., Introduction to Algebraic Geometry, to be published.

14. Hochschild, G., and Mostow, G. D., "Representations and representative functions of Lie groups", Annals of Math., 66 (1957), 495-542.

15. King, H., "Approximating submanifolds of real projective space by varieties", to appear.

16. Kuiper, N., "Algebraic equations for non-smoothable 8-manifolds", Publ. Math. IHES, 33 (1967), 139-155.

17. Lang, S., Algebra, Addison Wesley, Reading, Mass., 1965.

18. Lang, S., Introduction to Differentiable Manifolds, Interscience, New York, 1962.

19. Lang, S., Algebraic Geometry, Interscience, New York, 1968.

20. Macdonald, I. G., Algebraic Geometry; Introduction to Schemes, Benjamin, New York, 1968.

21. Milnor, J., Singular Points of Complex Hypersurfaces, Annals of Math. Studies, No. 61, Princeton University Press, Princeton, 1968.

22. Morrey, C. B., "The analytic embedding of abstract real analytic manifolds", Annals of Math., 68 (1958), 159-202.

23. Mumford, D., Introduction to Algebraic Geometry (preliminary multilith version).

24. Nagata, M., Local Rings, Interscience, New York, 1962.

25. Nash, J., "Real algebraic manifolds", Annals of Math., 56 (1952), 405-421.

26. Palais, R., "C^k actions of compact Lie groups on compact manifolds are equivalent to C^∞ actions", Amer. J. Math., XCII (1970), 748-760.

27. Shaferevich, I. R., Basic Algebraic Geometry, Springer-Verlag, New York, 1974.

28. Seifert, H., "Algebraische approximation von Mannigfaltigkeiten", Math. Zeit., 41 (1936), 1-17.

29. Tognoli, A., "Su una congettura di Nash", Annali di Sc. Norm. Pisa, 27 (1973), 167-185.

30. van der Waerden, B. L., Modern Algebra, Ungar, New York, 1949.

31. Wallace, A. H., "Algebraic approximation of manifolds", Proc. London Math. Soc., (3) 7 (1957), 196-210.

32. Whitney, H., "Elementary structure of real algebraic varieties", Annals of Math., 66 (1957), 545-556.

33. Whitney, H., "Analytic extensions of differentiable functions defined on closed sets", Trans. AMS, 36 (1936), 645-680.

34. Zariski, O., "Analytic irreducibility of normal varieties", Annals of Math., 49 (1948), 352-361.

35. Zariski, O., and Samuel, P., Commutative Algebra, van Nostrand, Princeton, 1958.

Randall Library – UNCW
QA612 .P295 NXWW
Palais / Real algebraic differential topology
304900276407%